Environmental Forensics
(Forensic Meteorology)

How the Atmosphere Affects Criminal Investigations
& Other Professional Research

By

Greg MacMaster
Forensic Meteorologist

Cyclogenesis Publishing
Publishers since 1992

Greg MacMaster   Environmental Forensics

Copyright © 2010-2020

All rights reserved except where previous work has been attributed. Printed in the United States of America. Some part of this publication may be reproduced, stored in a retrieval system, or transmitted, in any form or by any means electronic, mechanical, photocopying, recording or otherwise, with written permission of the author.

ISBN 978-0-578-76321-7
Publisher - Cyclogenesis Publishing

www.forensicweatherman.com

Greg MacMaster   Environmental Forensics

Environmental Forensics

Table of Contents

Purpose:	Environmental Forensics	7
Chapter 1	Forensic Meteorology	11
Chapter 2	Accessibility & Availability of Data	28
Chapter 3	Legal Understanding, the Basics,	43
Chapter 4	Pathology/Medical Examiner	78
Chapter 5	Entomology	104
Chapter 6	Forensic Toxicology	132
Chapter 7	National Guidelines for Death Investigations	137
Chapter 8	Aviation Accident Investigations	167
Chapter 9	Construction	184
Chapter 10	Dendrochronology – Tree Dating	192
Chapter 11	Auto Accidents	200
Chapter 12	Drowning Victims	222
Chapter 13	Hydrodynamics	242
Chapter 14	Ghost Ship	244
Chapter 15	Search & Recovery	265
Chapter 16	Sample case files from archives	272
Chapter 17	Resources, Links	278
	Bibliography	280

Greg MacMaster   Environmental Forensics

Book Reviews:

Forensics - the new buzz word. Finding a compilation of topics on this new and fascinating field is not easy. Greg MacMaster has created a terrific book that introduces many topics in the complex and varied field of forensics. This is a 'must read' for anyone who has an interest in this rapidly emerging area. It is required reading for the 'Forensic Meteorology' class that I teach at the University of Nevada at Reno and I am sure it will soon be required in many other forensic courses across the nation. A great introduction to a fascinating study!

Elizabeth J. Austin, Ph.D., C.C.M.
President
WeatherExtreme Ltd.

"As a consulting meteorologist, Mac Master's book has been extremely helpful in my preparation of reports for law firms and insurance companies. It's a great reference book for anyone interested in beginning work as a forensic meteorologist. The book acts like a "Forensic Meteorology Reference Guide", complete with tips on how to gather information and how to bill invoices to clients. It was exactly what I needed when I was getting my forensic meteorology business off the ground and I'm extremely grateful he wrote it. I can't wait for his next book."

Aaron Mentkowski, CBM
WKBW-TV Buffalo, New York, NY, USA
Meteorologist

"I have just gone through the Environmental Forensic book. I can say with some authority, this is an instant bestseller. Please have it published immediately. Forensic community would love to read it (besides of course people in the meteorology business.) Congratulations for publishing such a great work!"

Dr. Anil Aggrawal is a professor of Forensic Medicine at the Maulana Azad Medical College, New Delhi

"The influence of temperature is the most important one to the object of my forensic focus: insects on corpses. This book may help as a starter not only to check for temperature but also for microclimatic changes in forensic entomology cases. It also stresses that interdisciplinary methods – i.e., experts from different fields working together, not one expert trying to become an expert in too many fields – that leads to truly successful understanding of what happened when and how on a scene of crime."

Dr. Mark Benecke: Expert for Collection, Examination and scientific Interpretation of Biological stains from Crime Scenes. Certified Forensic Biologist, International Forensic research & consulting. Koln Germany.

Environmental Forensics

The purpose of this book is to equip the student with sufficient background knowledge of the following subjects and to appreciate their application to investigations and introduce the science of forensic meteorology and its effects on all types of investigations.

The information is either directly or indirectly related to the science of meteorology and serves as a tool to understand what investigators are up against when conducting investigations and gives the meteorologist a keener sense of awareness when assisting in the investigation. It is not intended to be a [beginner's guide], but more of a source of information about the range of expert knowledge that is available to assist in investigations and create a starting point to make the investigator an intelligent client of the expert. The nice part about this manual is that it brings all respected professionals together under one umbrella and galvanizes their expertise into a 'successful team environment'.

Forensic Meteorology (a division of atmospheric forensics) is the science involving the application of meteorological analysis to cases involving events for which Medico-legal, legal and/or insurance claims are being made. The forensic meteorologist (F.M.) "reconstructs" the weather near the time and place of the event, using and interpreting all available data sources, including weather observations, weather radar, and satellite, lightning detection equipment and any other information that deem necessary to the case. The F.M. will need to expand on his/her technique by performing single station recordings where comparisons of weather elements are needed. That means visiting the crime scene!

"What is a Forensic Meteorologist" is something I get asked quite often and the image of the broadcast meteorologist on TV is what most people envision. After an explanation of what a forensic meteorologist is and what they do, they think CSI the hit TV series, and to a degree I can relate our profession to some shows where the weather played a major factor in deciding the outcome in someone's fate, although certainly not as glamorous. There are clues in every case and weather is one of them. But it's not just how the weather is reported and reviewed but more as to [how] the weather affects other elements within the case. Convection, conduction and absorption of the suns rays can alter the "traditional" temperature readings. This, in turn, can help narrow down a timeframe. It can be a cause of an accident and it could even remove someone who was named a prime suspect. Wind blowing over a lake will push surface water in one direction and in the opposite direction near the bottom of the lake. This is key when searching for a drowning victim.

All available data sources can and should include non-traditional methods of weather recording from armchair weather enthusiasts, local farmers, Schools, Department of Transportation records or anything that can lend a factor in the analysis. The elements of weather have a dramatic influence not only in our future but also in our past. Horticultural Research Stations are extremely valuable in data (surface, ground and sub-terrain recordings) that would normally be non-accessible due to limiting sensor data.

There is no "distinct" definition of "Forensic Meteorology" that is analogous among publications and opinions widely vary.
In its simplest form, Forensic Meteorology pertains to the law or legal system. Some colleagues I conferred with agreed. So what follows is what was presented at the American Meteorological Society Forensic Workshop, San Diego, CA. 2005 by Sean Potter, CCM.

Forensic Meteorology as defined by the World English Dictionary, (North American Edition)

Forensic: From the Latin word forensic, meaning "of the forum", relating to the application of science to decide questions arising from crime or litigation.

Forensic Meteorology: The application of meteorological and climatological data to legal cases and related investigations in which weather may have been a factor.

The role of the Forensic Meteorologist: "To assist the court in the search for the truth by presenting the most accurate description possible of the meteorological events pertinent to the case of litigation." (Haggard–2003)

"To fulfill this role, the forensic meteorologist must utilize the best possible data set in his or her retrospective reconstruction of the weather. The quality of the analysis is dependent on the availability of representative measurements of meteorological parameters." (Haggard–2003)
(As presented by Sean Potter, CCM, at the AMS Conference Workshop for Forensic Meteorologist, San Diego 2005)

The forensic meteorologist, who may act as either a background consultant or a testifying expert, will collect, interpret and analyze atmospheric data in support of insurance fraud claim investigations, civil and criminal trials, and environmental regulatory actions. The forensic meteorologist may be employed directly by an insurance company, the attorneys for either the plaintiff or defendant in a case or, with increasingly frequency, may be appointed by the court itself. Regardless of the employing party, it is not the role of the meteorologist to be an advocate for either side in a dispute, but to assist the judge and/or jury in understanding the often complex facts in a case so that they may reach an appropriate verdict.

The legal definition of Expert Witness: When knowledge of a technical subject matter might be helpful to a trier of fact, a person having special training or experience in that technical field, one who is called an expert witness, is permitted to state his or her opinion concerning those technical matters even though he or she was not present at the event.

Some typical problems we (expert witnesses) deal with in forensic meteorology are pretty typical and range from the following:

"The automobile accident was caused by poor visibility - was that caused by natural fog or pollutants from a nearby industrial plant?" "Was the building damaged by a tornado or straight line thunderstorm winds?" "A person was found electrocuted near a downed power line - was it a fault in the utilities' line or a lightning strike?" "How can we demonstrate that rain fell at a site that is located many miles distant from any National Weather Service reporting station?"

Greg MacMaster Environmental Forensics

The forensic meteorologist may collect standard and non-standard weather observations, assemble weather radar and satellite imagery, process weather data taken by a party in the case, or locate nonstandard sources of data such as lightning ground strike reports or atmospheric data taken by air pollution monitoring networks. This accumulated data is then used in a comprehensive analysis of the meteorological facts pertinent to the case. There is increasing use of sophisticated computer graphics and video animation of weather information in trials and mediation/administrative hearings. Most forensic meteorologist has had long and varied careers in the atmospheric sciences, and it is their hard-earned expertise that is in demand. Most successful forensic meteorologists have met the qualifications of a Certified Consulting Meteorologist (CCM).

Now that we have a fairly good idea on the definition and role of the forensic meteorologist, but how [does] it affect other forensic scientists in their field of study? How much of an influence does our research have on the Entomologist (Study of insects), Engineer, Construction Company (residential and commercial), Transportation (aviation, automotive, rail and marine), Pathologist (Medical Examiner) and Law Enforcement? Let's find out!

Greg MacMaster   Environmental Forensics

Chapter 1

Forensic Meteorology

How Can Forensic Meteorology Impact an Investigation?

It was mid-afternoon Monday (3 P.M.), cool and dry with a mix of sun and clouds when the groundskeeper heard 2 dogs, chained up and barking uncontrollably towards the house. He walked over to investigate why the dogs were barking and, peeking through the glass noticing a body lying on the floor face down, he dialed 911 from his cell phone and Police and Emergency personnel soon arrived to determined that due to the advanced rigor mortis state - he had been dead for quite sometime. They radioed for the Medical Examiner and after an hour, officially pronounced him dead at (4:30 P.M.). He then took a body core temperature. It was 86 degrees.

The room temperature was reported to be 75 degrees, however, the electronic thermometer changes at 11 p.m. to 70 degrees. He was found on the main floor, there's also a basement and second floor. The house is immaculate, clean to the touch with no dust anywhere. It was obvious he was a neat person. The house was heated with baseboard heat and all windows were closed and locked, except for the door to the backyard. As a precaution, the M.E. took an outside temperature of 62 degrees at 4:45 P.M. Light winds and partly sunny.

The extremities of the deceased were fixed, hard to move. He was wearing light clothes and for the exception of a small cut on his face, which revealed after they rolled him over, indicated Blow Fly eggs had been deposited. After a cursory inspection of the grounds around the house, they found nothing highly unusual except for the open shed door in the back with a few items out of place. They did notice some marks which indicated the dogs dragged a large stick, which was close by, across the yard.

In the course of the investigation, the local meteorologist gets a call and is asked to provide the local weather conditions for that day in question. He obliges and gave the high and low temperature for the day, .16" of rainfall and that winds were gusty in the early morning hours. But was that enough for the investigators to make a determination on estimated time of death? Was it a natural death, homicide or murder?

That's not for a forensic meteorologist to decide. But having knowledge in other facets in forensics would give clues as to what may have happened and in the process, give us a better idea as to what to provide the investigators. The deceased was a neat person, living clean, yet fly eggs were found in a facial cut that was against the floor? Drag marks in the yard suggesting the dogs may have played around (yet they were chained up). Questions start to surface to whether he was killed outside, and then moved indoors to hide the crime. He may have been outside long enough to attract flies and then was moved indoors. That would explain the fly eggs and drag marks. What about body temperature? The body cools at an average of 1 to 1.5 degrees per hour. Errors in comparison readings are probable without having knowledge in instrumentation and recording guidelines.

If you had a working knowledge in how a body cools and estimating time of death, you would find a conflict in some findings.

1. The body temperature indicates that he would have been dead over 24 hours (if compared to the temperature inside the house).
2. Using outside air temperature (62 degrees) in the formula, you find that the estimated Time of Death, (ETOD) would put it at 13.2 hours to 18.8 hours (between 9:42 P.M. the night before to 3:18 A.M. the following morning). But as a forensic meteorologist you notice something in greater detail that could decrease the timing error.
3. The M.E. recorded a temperature, but at what height above the ground (or was it on the ground)?
4. Meteorologists are aware of how winds make us feel cooler and know that water cools the body 3 times faster. Formulas can prove this theory. So what do you do?

A study in my own house on temperature -
Thermostat is located on main floor (5 feet high)

Second story bedroom	70.7	on the bed
Main floor wall	69.7	5' off floor

Placed on carpet /main floor   67.9
Basement floor   65.7   on carpet

There's a 5 degree difference between the basement floor and the second story bed. Would this 5 degree difference change the ETOD? If a body was found in the second story bed, would the M.E. use the thermostat reading for an outside air temperature? If inside temperatures vary this much, how much of an impact would a change in outside temperatures have?

Education, skill, working knowledge and experience, by far, are the most important factor in determining if you want to be an expert witness. Simply researching archived weather data isn't enough to label you a forensic meteorologist (FM). It's how that data is applied to the research and findings. To search the past correctly, you have to know how to forecast.

Knowledge in forecast models, upper air soundings, surface observations and other areas help you understand the fluid dynamics of how the atmosphere works. This will heighten your visual perception when you appear at a crime scene and have to find topographical influences that may have been overlooked by the accused, their legal representation or other professionals that may have been called as "expert witnesses".

With that said, you should have a minimum of educational requirements to enter into the Forensic Meteorologist profession. A degree in the Atmospheric Science field, preferably Meteorology is preferred. Other fields of discipline within the atmospheric science field may be;

Applied Climatology
Meteorology as it applies to Aviation, Marine
Mountain wave
Solar
Oceanography
Satellite Meteorology

Radar Meteorology
Thermodynamics and Physics

Applied Meteorology
Instrumentation
Dynamics

It would be a good idea to have an understanding of atmospheric processes on all scales. Above all, know your professional and personal limitations. Lack of knowledge can hurt the case, your credibility and your future. Additionally - courses in business (marketing, accounting and business ethics) as well as legal courses would strengthen your knowledge in your forensic business. If you want to expand your knowledge in other areas, such as, medical, building construction, law enforcement, aviation and so on, take classes and work in that environment and see how weather plays into that science.

You can never get enough education. Most Government and Private Sector employers offer in-house training. Seasonal workshops are becoming more prevalent in the National Weather Service and often use analysis techniques to determine what caused a/an excessive rainfall/snowfall, severe weather event. This type of training is invaluable! There is a number of web & computer based courses to help you gain knowledge in areas where you feel you may be lacking in knowledge. Scientific literature in your specific field of study will enhance your technical knowledge. The American Meteorological Society is the foremost authority in education criteria with established scientific manuals designed to assist you in this profession.

Experience: How does one go about getting experience? Well, don't expect to be a forensic meteorologist right out of school. It takes years of on-the-job training to develop skills necessary in the research field. You learn by your verified forecast and busts, more commonly termed (successes and failures.)

A way to gain experience:
1. Professional experience in applied research and consulting,
2. Teaching experience – You will be called upon to instruct legal counsel, judge, jury or other professionals on

   information pertaining to your findings. This is not the time to be nervous, so you'll need to be comfortable
3. Private, Governmental and Corporate world

4. Years of (non-forensic) work experience
5. Working under an expert witness helps shape the mindset in forensics
6. Review your forecast and research conditions that changed which to a missed forecast.

Experience also helps you expand your knowledge in how the weather impacts our daily lives. Simply forecasting isn't enough, but how that forecast impacts our pattern of living day to day, week, weekends, shopping habits, travel itinerary, vacations, farming, harvesting. I could go on forever. This is how you develop a better appreciation how sensitive we are to the weather. This sensitivity helps us to find the truth in research.

Other ways you can educate yourself on other professional fronts: Enroll in classes designed for that specific profession at a local college or University. In the EMS/Fire field, classes are held annually to maintain their credentials. There are many ways to increase your knowledge in other professions. But it takes drive and ambition.

So what do we do as forensic meteorologists? (See Chapter 3 on the specific steps of protocol). Well, to start, we acquire data and interpret it for those who are unfamiliar to the "weather lingo". We have to simplify the terminology so it's easily understandable in a court of law. Keep in mind that you may have to present your finding in front of a jury.

These people are from all walks of life, ethnicity, income and professional status. Your jury may consist of a baby-sitter/Nanny, Heart surgeon, Boat Capitan, College student, Secretary, retired, School teacher…You get the idea. This is where your ability to teach/present comes into play. You also will advise clients on the importance of the weather and how it may have played a factor in the case. You may have to perform a multitude of specialized analyses in order to find a variable to assist in your findings. (Upper air dynamics, thermodynamics, micro-analysis, etc)

One of my more popular cases which defined my credibility in forensic meteorology came with a call from a lawyer asking what the wind data was for a specific day. This data came from a local airport and the site of the house fire was miles away (as the crow flies). The airport, which lies in a valley oriented north – south, showed a wind out of the South at 7-9 mph. When I learned more about the case and the questions posed by the attorney, I found that I would have to do a micro analysis from the upper atmosphere to determine the wind flow at the surface. The regional wind plots showed varying directions and speeds and the accident (2 houses caught on fire) occurred on the East shore of Lake Michigan.

To make a long story short, had I used the airport data (single source); one insurance company would have saved millions of dollars in payout. Instead, the data and supporting evidence I had to substantiate my analysis, along with formulas and operational techniques, allowed me to come to the conclusion that the wind flow was from the west. This shifted the responsibility to the other insurance company for pay-out. But the legal ramifications could have been much worse. Had I presented the wrong conclusion, fault could have fallen on the builder. The future of a family business was hanging on the testimony.

Evaluating a report from the opposing weather expert witness (or other witnesses) may shed light on a situation which could help strengthen your case. I have found it best to prepare my expert witness draft report before I review the opposing legal testimony. Review with your legal counsel and then finalize your draft for submission. At the end of this book I have a sample paragraph which will leave options open to submit additional information at a later date.

As you can see, experience comes from years of working in the field and being subjected to the daily routine of your profession. Don't expect to be a success or get rich overnight. It's a great second income and should you decide to venture into a full-time business, expect allot of slow/down times. Another words, have another source of income until you become well established.

There are thousands of meteorologists Nationwide employed by corporations, governmental sectors, broadcast stations and self employed. How do you know if you have the qualities to be a good forensic meteorologist? Well, for starters, you have to have the desire to find the truth through tedious hours of research for little or no pay.

If you're getting into the business for money, you're getting in for the wrong reasons. If you're a big fan of CSI, Forensics and Legal / Court Files, then your have a running start.

If you would like to test your skills and see if you have the fortitude, try interning with a forensic meteorologist. Accompany him/her to accident sites and ask questions (but don't get in the way). Review previous cases with them.

Here's a site where you can share your thoughts/questions or if you're just looking for some insight. http://groups.yahoo.com/group/forensic_meteorology/

Finally, there's respect for our fellow forensic meteorologist, as it's not uncommon to have an expert witness on both sides (Plaintiff and Defendant). We can't share ideas and thoughts so we work independently and for good reason (conflict of interest). Keep comments about the opposing testimony professional and respectful. Remember, we all have the right to make modifications and changes should we become aware of additional information that could sway our testimony. If you're unfamiliar with an aspect that you feel would help your testimony, seek the advice of a fellow forensic meteorologist (not associated with the case) who may have better insight to that specific science, (Air pollution and chemical dispersion) as an example. Be prepared to offer up a portion of your pay to offset the cost for information. Remember, they're in the business to make money too.

Private-Sector Meteorological Consulting: View through the Lens of a Sole Practitioner

Phillip D. Falconer, C.C.M.
Falconer Weather Information Service
Scotia, New York

Invited Presentation

"Vision Seminar Series"
National Weather Service Headquarters
Silver Springs, Maryland
November 8, 2000

Greg MacMaster   Environmental Forensics

I am gratified that Gen. Kelly and the organizers of the National Weather Service's Vision Seminar Series have invited me to appear before you, to represent the views of another of your constituents from the ranks of the private sector, namely that of the meteorological consultant. As much as I might wish to consider myself a visionary, I feel best qualified to share with you a few thoughts I have on the present-day health of the weather consulting business, and touch on a few challenges and opportunities I see for this segment of the private sector in the coming few years.

I know that you have previously heard from representatives of some of the country's largest and most prestigious operational forecasting firms and value-added weather information suppliers; their views on the appropriate roles of the National Weather Service and of the commercial weather industry are important to hear and discuss. However, the business of weather in the United States is diverse, and the views of the commercial weather industry are not necessarily joined by each of the various constituent groups comprising the private sector meteorological community. Indeed, no one segment of the private sector can lay claim to a unified view of the appropriate roles and responsibilities of the National Weather Service and its private sector partners.

For many meteorological consultants, most of whom are not directly engaged in operational forecasting, recent discussions regarding future role of the National Weather Service in the provision of our country's weather and climate services, while evocative, have seemed at times both remote and unsettling. With interest in the topic running high, it was not altogether surprising that, in July 1999, the American Meteorological Society adopted a Policy Statement on the Public/Private Sector Partnership, expressing its view on the appropriate roles and shared responsibilities of the government and of the private sector in the provision of this Nation's weather, climate, hydrologic and environmental services.

This policy statement is, in my opinion, a balanced and reasoned view of the core issues surrounding this partnership, and was developed in conjunction with the nine-member AMS Board of Private Sector Meteorologists.

The Business of Weather Consulting Comes of Age

Recent estimates, although rather uncertain, suggest that there are perhaps 300 to 400 'for-profit' firms nationwide, engaged in the business of weather, and ranging in size from individual consultants to corporations employing dozens of meteorologists and enabling support technologists.

The private sector meteorological community is engaged in a stunning array of service initiatives, including, but certainly not limited to the sale and delivery of basic weather and climate data; the creation of value-added meteorological data products and enhancement of weather data visualization capabilities; operational weather forecasting; meteorological and environmental consulting services; and the production of weather monitoring, communications and computing technologies. Many private meteorologists work directly for weather-sensitive industries, including the gas and electric utilities sectors; the airline, transportation, shipping, and agribusiness communities; the broadcast media; and in commodity exchanges.

While there is a rich history surrounding the humble beginnings and gradual development of the commercial/industrial weather sector over the past half century, the pace of growth in the private sector, and the applied technological innovations it has contributed to the science of meteorology, has noticeably accelerated during the past decade. As we enter the 21st century, roughly 1 out of every 3 meteorologists has found employment in the private sector. By the year 2005, the number is expected to swell to 40 percent. And, by the end of this decade, nearly half the nation's meteorological work force will likely be privately employed. In the competitive marketplace, this growth will undoubtedly spur a host of changes in the manner, efficiency and costs of delivering applied, customer-specific weather information to the Nation.

'Boot Camp' for the Forensic Meteorological Consultant

Much of the meteorological consulting work I have done since 1980 or so involves the acquisition and analysis of past weather records for insurance claims investigations, and matters of tort litigation. Those who have been involved in the forensic meteorology specialization have seen this segment of the profession grow steadily over the past twenty years. Several dozen firms, many of which operate as sole proprietorships, now offer forensic meteorological services; some are able to do this profitably on a full-time basis.

Other than three American Meteorological Society workshops on the application of weather information in legal proceedings, held between 1989 and 1996, and a few generalized primers on the presentation of scientific evidence in court, the meteorologist is hard-pressed to find any opportunities to learn about the intricacies of the U.S. judicial system and the manner in which expert weather witnesses are expected to participate in this system. This points out, of course, a need which is likely to be filled in the next year or two. Presently under discussion at the National Council of Industrial Meteorologists is how we might facilitate one or more workshops, where meteorologists would be able to fine tune their courtroom skills in a mock trial setting.

Participants would be introduced to a hypothetical law suit involving some aspect of the weather; would watch the matter litigated; and would then have the opportunity of evaluating the proceedings. Trials would be held in a mock courtroom, with a presiding trial judge, opposing attorneys and meteorological experts, and demonstrative exhibits. Such a continuing educational program would touch on each of the key elements of forensic weather investigations:
- pre-trial depositions and expert affidavits;
- courtroom rules and procedures affecting expert witness testimony;
- the introduction and use of meteorological evidence at trial;
- the objectives of direct examination and cross-examination of the weather expert; and
- the role of the weather expert as an educator in the courtroom.

There has been talk about facilitating such seminar opportunities for a few years now, especially in view of the fact that previous workshops were largely a pot pourri of meteorological case studies and 'war stories', often with limited discussion of the key legal aspects of the case; of the lawyer-expert deliberations prior to and during trial; the manner in which demonstrative meteorological evidence was used, or the scope and general substance of opposing counsel's cross-examination tactics. Any one of these topics would provide useful guidance to those generally unfamiliar with the 'drama' of litigation.

The time has come for our profession to encourage 'hands-on' training opportunities for those meteorologists who find themselves increasingly involved with weather-related insurance claims investigations and expert testimony. This means collaborating with representatives from the claims industry and legal profession, not only to insure that meteorologists understand what is expected of expert witnesses, but for the insurance industry and attorneys:

- to understand how, when and where weather information is collected;
- how it is disseminated to the public;
- what the various weather data products mean; and
- why the selected records bear particular relevance to the case being litigated.

Harvesting the Nation's Weather Data Resources: Positive Steps-Ongoing Challenges

Continuing this line of thought, I wanted to touch on an issue that bears obvious impact on the meteorological consultant's ability to document past weather conditions, based on the present-day ensemble of NWS data products. As a result of the agency's multi-year modernization program and as generally set forth in its recent Strategic Plan for Weather, Water and Climate Services through 2005, Weather Service customers and partners now enjoy convenient and direct internet access to a wide array of operational data sets, charts, images, and technical reports. Not too long ago, many of these products were available, by request, only in hardcopy format through the archives of the National Climate Data Center (NCDC) or through third-party commercial weather information service providers; this

proved to be a costly, often time-consuming expense for meteorological consultants, particularly in situations where their clients required expedited service. More open access to a wider variety of Weather Service documents and products has allowed many meteorological consultants to build their own digital libraries of weather, water, and climate information, thereby enabling them to more efficiently pass along operational data and value-added products to their customers.

However, with as many benefits as have arisen out of the NWS modernization efforts, the reality for many meteorological consultants is that today's easy access to more National Weather Service data products has been offset by some operational and reporting limitations in both the national network of automated Surface Weather Observations systems and in the Hourly Precipitation Data station network.

Over the course of the past five years, those of us who routinely use surface weather observations in reconstructing past weather events are painfully aware that, while there are now more reporting stations across the country, some providing more frequent observations than ever before, the loss of, for example, the six-hourly synoptic reports at the automated surface observing system (ASOS) stations, particularly snowfall and snow depths; the lack of thunderstorm reports at many stations; and the replacement of nearly all of the Universal weighing rain-gages with (lower-resolution) Fisher-Porter-type gages across the nation's rain monitoring network have had the unfortunate and unintended consequence of jeopardizing the weather analyst's ability to fully-evaluate past weather conditions.

It is not my intention to go into a litany of concerns that have been expressed over the operating characteristics of automated weather observing systems, or of the Fisher-Porter rain gauge. But institutional changes in this country's basic weather observing infrastructure has, and will likely continue to affect those of us who rely so heavily on National Weather Service data products, particularly in the 'snow and ice' belts of the United States, where we often have to make do using proxy indicators and supplemental private sources of weather information.

This does not go unnoticed by our clients, in part because the conscientious weather consultant may need to advise them of any known limitations in the basic National Weather Service measurements or reporting procedures, which might compromise an expert reconstruction of past weather conditions. Again, I mention these matters not so much as a criticism of our present-day national weather observing infrastructure, but as an opportunity to provide you with feedback from a private sector constituent.

Desktop Computing Power Unleashes New Opportunities for Weather Consultants

The arrival of fast, affordable, powerful desktop computers; of wide arrays of digital data sets; and of gigabyte-sized data storage devices has enabled meteorological researchers and modelers to quickly retrieve and process the large volumes of data needed to run specialized weather forecast, atmospheric flow simulation, and air pollution transport and dispersion models. These technology developments have also allowed meteorologists to archive vast amounts of historical weather data and data images in digital format, which may then be easily retrieved for application to specific case studies, or to broader climatological or statistical analyses. With these tools, the consultant is able to offer customers a broader and more sophisticated selection of work products than were ever before possible, often on a faster turn-around schedule.

The data ingest and processing speeds necessary to view animations of Doppler weather radar, high resolution satellite images, and lightning ground strikes has been well-within the capabilities of desktop computers for many years. More recently, sophisticated, computationally-intensive predictive and diagnostic models of the atmosphere have been developed, which now can run on stand-alone PCs. One Minnesota-based commercial firm uses a PC-based version of the NCAR-Penn State University MM5 mesoscale meteorological modeling system, to generate real-time weather forecasts as well as three-dimensional flow visualization and pollutant transport simulations in complex terrain. Another forecast consultant now provides sailors and outdoor adventurers with customized wind and weather forecasts, based on the ETA model, which he runs on a Linux PC workstation. A Georgia-based company markets a

Windows-based hurricane tracking, analysis, and planning software program, with numerous capabilities tailored for the emergency management community, including SLOSH storm surge data; a SLOSH inundation level analysis subroutine; and various user-selectable impact assessment tools.

Soon analysts will be able to routinely explore geographically-referenced environmental data sets; to test various 'what if' storm threat scenarios; and to search for and evaluate spatial relationships between meteorological and environmental properties using complex queries. Work is underway within NOAA's National Data Centers, especially development of the NOAA Virtual Data System, which employs new GIS technologies and GIS-enabled databases, will facilitate online, browser-based information mining.

It is not only likely that opportunities will arise for the private sector meteorological community to partner with government and academic institutions in developing the visualization tools and underlying geo-referenced environmental data sets, but that commercial weather enterprises will also enter this market on their own, capitalizing their risk entirely with monies derived from the private marketplace.

Minding Our Business in the Classroom: A Plea to Develop University-Level Opportunities Relevant to the Commercial Weather Service Industries

I mentioned earlier there should be ongoing opportunities for meteorologists working in the forensics area to improve and sharpen their skills, by participating in focused workshop practica designed around the concept of the 'mock trial'. But there is fundamentally a much broader need for specialized education related to the private practice of meteorology. Undergraduate atmospheric sciences curricula at universities across the nation simply have not done a particularly effective job in teaching students about the business aspects of applied meteorology. Many meteorology students today will, upon graduation, work in the private sector; it is disappointing to report that most will not have had the opportunity to learn the fundamentals of sound business practice, of customer service, or even how weather affects industry operations and its infrastructure.

While unsettling, this is not particularly surprising, as relatively few atmospheric science faculty have first-hand, private sector experience in operational meteorology or with the business aspects of full-time meteorological consulting. The University of Oklahoma at Norman has tackled this issue head on. Recently, the School of Meteorology and College of Business have collaborated to offer an undergraduate Area of Concentration in private sector meteorology. WeatherData, Inc. of Wichita, Kansas has also attempted to address the issue, by organizing a 3-1/2 day workshop, "Introduction to Commercial Meteorology for University Faculty", designed to illustrate how education centered on real world, workplace-oriented challenges is important to success in the private weather industry.

Faculty workshop participants partner with company meteorologists, who review how weather affects their clients, and guide participants through the process of preparing client-specific weather forecasts on short deadlines, alerting their customers to pending severe weather, and handling research requests. This is precisely the nature and scope of effort needed in our universities to prepare the next generation of meteorologists not only for work in the commercial weather industry, but also across the spectrum of private sector meteorological opportunities, including the business of weather consulting.

Some Thoughts on National Weather Service Participation in the AMS Certified Consulting Meteorologist Program

Akin to the licensing of doctors, attorneys, accountants, engineers and other professionals, the Certified Consulting Meteorologist (CCM) program of the American Meteorology Society assures the public that the individual possessing the CCM designation has successfully completed a rigorous peer-testing evaluation, and has met or exceeded certain high standards of technical competence, character and experience. Since its inception in 1957, the AMS has issued over 600 certificates, most of these granted to private sector consultants, industrial meteorologists, and atmospheric scientists from the university community and, to a lesser extent, to meteorologists employed in state and federal agencies and government-supported research institutions.

Relatively few National Weather Service meteorologists have ever applied for the CCM designation. Why this is so is unknown, but may relate to any one or more of the following reasons:
- confusion over the meaning of the term 'consulting meteorologist' and whether certification as a consultant would be perceived to be in conflict with, or otherwise

jeopardize the government meteorologists' traditional public service function;
- the fact that the NWS already administers several internal continuing education and training programs leading to professional qualification;
- that relatively few NWS field staff are members of the American Meteorological Society; or
- that the Society's marketing efforts for the CCM program have simply not been effective.

Whatever the reason, there continues to be few National Weather Service certificate holders.

You may ask, then, why should I, as a private sector consultant, be at all interested in raising the visibility of this particular professional certification program within the Weather Service? I believe the CCM designation represents a fair and reasonable opportunity for those in both the private and public meteorological communities, those who in one fashion or another 'consult' with their constituents, to demonstrate a common minimum standard of technical excellence; parity, if you will.

The current AMS Certification Board includes representation from the National Weather Service, and efforts to attract operational meteorologists are under active consideration. Many CCMs have voiced their belief that the Board testing procedures could be tailored to a candidate's area(s) of technical competence, including operational forecasting and weather analysis. In my opinion, the pool of examination questions and case studies for operational meteorologists seeking the CCM designation, should be updated and expanded, possibly in coordination with the National Weather Service training facilities in Norman and Kansas City and with the COMET Program Office.

I believe that where the professional interests and opportunities for technical achievement in the private and public sector meteorological communities intersect, a healthy mutual understanding of each other's abilities would emerge. The CCM program may represent just such an opportunity. Whether it deserves a second look by National Weather Service meteorologists, who may not have previously considered its relevance to their time in service, or who may have felt it merely a program for private meteorologists, remains to be seen.

Final Thought

Permit me to leave you with a final thought, that perhaps consideration be given to a simple, low-keyed program through which National Weather Service meteorologists, particularly those in your field offices, could be given leave time to shadow the activities of their private sector colleagues in the weather consulting business. Observation at close range of the consultant's daily workload, from the intake of client requests, to the acquisition and analysis of National Weather Service data records, and eventually to the preparation of value-added meteorological data products and reports, would go a long way in fostering an understanding and appreciation of the meteorological consultant's role in the business of weather in this country.

Chapter 2

ACCESSIBILITY & AVAILABILITY OF WEATHER DATA

The information super highway has opened up a whole new area in accessing weather information. Hardcopy records are still available at many local libraries, colleges and universities and some records are kept at the local level in governmental centers. This list is endless and it's up to you to be creative and think outside the envelope. Don't trap yourself with limited data. The more you have available, the better your decision making process.

First and foremost you will need official records, and to do this, you'll need access to the NCDC website. I would suggest you sign up for their newsletter and make sure they have your name for future enhancements and updates. It's a fee-based system, and charges vary depending on amount of data requested. You could use a more

indirect means of access data for a preliminary review by going to www.weatherunderground.com and go to the history link. It's a quick view of what you could expect to encounter but shouldn't be relied upon as "Official Data". Some data may not be available, so try other searches for specific weather information. Some search engines used as of this writing (Google, Hotbot & yahoo). Check with your local National Weather Service Office and see what they keep on record and how long they keep it on file before they send it to NCDC for archiving.

LEGAL, INSURANCE, and CERTIFICATION
Information for legal and insurance claims
Information on data certifications.

What is "certified" data? Most states require all records that are to be submitted as evidence in a court of law to be authenticated in some way, typically called certification. There are several types of certification available. The National Climatic Data Center, the official United States archive for climatic records, can provide more comprehensive types of certification.

What can you testify to in court (and where can you testify)?
Generally, it is not necessary for expert witnesses to appear in court to

testify about the data we supply. If expert testimony is needed, the services of a forensic meteorologist should be retained. The American Meteorological Society and the National Weather Association maintain listings of consulting meteorologists.

Many outdoor events coordinators are looking to insurance companies for rain insurance. The threat of rain, in a measurable amount (usually .25" to 1.5) in a short period of time would insure a payout that would cover the loss of sales and/or damage to goods. Having a consulting meteorologist insures that the insurance company is getting accurate information that would otherwise be altered for financial gain. From outdoor concerts to antique auto shows, the amount of money to be lost from inaccurate reporting could be insurmountable to the insurance companies. So accurate reporting is a must.

An inspector said that there was hail damage to the roof of a home. "Can you tell me all of the dates that hail was reported in my area?" Well, that depends. First of all, hail is typically localized phenomenon. It may hail in one location, and a half mile away no hail may be observed. Because it can occur in a relatively small area, not all hail that falls is observed or reported. One of the criteria for a severe thunderstorm is observed hail ¾ inch or more in diameter, so if severe storms have occurred at your location there may be a record of hail if it occurred. These reports include the size of the hail, location where it occurred, and time it occurred.

For events less than three months old, check the preliminary daily storm reports on the Storm Prediction Center web. For events greater than three months old, check the on-line Storm Data database maintained by the National Climatic Data Center.

Official publications of Storm Data are compiled and published by NCDC. There is about a six-month lag in the publication of the monthly Storm Data publications. The minimum cost for receiving copies of Storm Data pages is $12.00 at the date of this publication.

If lightning data is requested, Global Atmospherics, Inc. is the place to go. They operate the National Lightning Detection Network, and provide custom reports on lightning strikes, and can locate strikes usually to within 500 meters. The NWS may be able to tell you whether or not a thunderstorm was reported in your area, depending on the availability nearby reporting stations.

What if you had wind damage to my roof during a recent thunderstorm, and the insurance company needs documentation of the wind speed before it will process the claim. Can you tell what the wind gusts were during the storm? Wind speed can vary greatly over short distances, and is affected by such things as trees, buildings, and other obstacles. Most wind speed and direction data is obtained from airport observations sites. Usually these have an unobstructed exposure to the wind and may not accurately represent the wind conditions at, for example, an urban location. If an airport observing station is located near your home, this data may be available and be sufficient evidence. How applicable this data may be to your location depends on your proximity to the observation site. If there is no wind observation site nearby, we would have to rely on the Local Storm Reports gathered by the National Weather Service or the Storm Data publication published by NCDC. Local Storm Reports are listings of storm-related tornado, hail, and wind gust/wind damage events. Wind speeds reported are often estimates and usually not at "official" observing sites. However, these reports may provide evidence that a severe storm occurred in your area.
(Information or portions therein provided by H. Berg, Midwest Regional Climate Center)

At the end of this publication, you'll find a plethora of links that an assist in your search for weather records.
Sometimes, data will be in raw form (numbers...and lots of them) and to decipher that data you'll need a computer program to crunch that data to give a more meaningful, simplified output. There's a plethora of programs to plot information and I'm sure I'll have to dedicate a chapter on this topic alone. We'll put this in the next revision of this book. I did a quick search and found a few (as of March/2005) which are referenced below:

http://grlevelx.com : A complete radar display / analysis program, allows for viewing Level II (hi-res) and Level III (NIDS) radar data in real-time or from archives.

www.raob.com : A powerful, multi-functional sounding analysis program that can read numerous types of rawinsonde/radiosonde data; create a variety of sounding diagrams; 3-D hodograms; time & distance based vertical cross-sections; mountain (lee) wave turbulence; and display over 150 atmospheric parameters including icing, turbulence, wind shear, inversions -- plus use of a unique severe weather analysis table, a cloud parameter table, and interactive diagrams.

www.wdtb.noaa.gov/resources/projects/BUFKIT/bufkit4.html : BUFKIT is a forecast profile visualization and analysis tool kit developed by the staff at the National Weather Service (NWS) office in Buffalo and the Warning Decision Training Branch (WDTB) in Norman, OK. BUFKIT is used in the public, private, and educational

meteorological sectors in both the US and Canada. BUFKIT is available to anyone who is interested in the analysis of forecast hourly profiles.

www.digitalatmosphere.com : Digital Atmosphere is powerful weather forecasting software that allows you to create detailed maps of real-time weather for anywhere in the world. It makes extensive use of techniques and algorithms that are comparable and in some cases superior to the National Weather Service's multimillion-dollar Advanced Weather Interactive Processing System. It can run on any 486 or Pentium system, and does not require any type of subscription or recurring costs, using free data from the Internet.

Archiving weather records is a cumbersome project and can take up tons of space on your computer and shelf space. So in lieu of the space problem, I have added a few links to assist in your research.

The U.S. Signal Office began publication of the maps as the War Department maps on Jan. 1, 1871. When the government transferred control of the weather service to the newly-created Weather Bureau in 1891 the title changed to the Department of Agriculture weather map. In 1913 the title became simply Daily weather map. Eventually, in

1969, the Weather Bureau began publishing a weekly compilation of the daily maps with the title Daily weather maps (Weekly series).

This site offers maps from 2002 to the present.
http://www.hpc.ncep.noaa.gov/dailywxmap/frame.html

For maps dating from 2001 to 1871, go to this site: http://docs.lib.noaa.gov/rescue/dwm/data_rescue_daily_weather_maps.html

(The following information was referenced, in part from the NCDC website and the NWS)

NCDC is the world's largest active archive of weather data. NCDC produces numerous climate publications and responds to data requests from all over the world. NCDC operates the World Data Center for Meteorology which is co-located at NCDC in Asheville, North Carolina, and the World Data Center for Paleoclimatology which is located in Boulder, Colorado.

http://lwf.ncdc.noaa.gov/oa/ncdc.html

You can link directly to an NCDC Weather Observation Station Record for a particular station (or list of stations) using the identifiers. Any call that finds more than one record will display a list of the records found. The user can then select the desired station from the list. To find only one record you should use Call Sign or WMO#. Follow this link to find out more:
http://www.ncdc.noaa.gov/oa/climate/linktowcs.html .

Certification information

The National Climatic Data Center (NCDC) is designated by Public Law 754, passed by the 81st Congress as the Federal Records Act of

1950, as the official United States archive for climatic data records. As an archive facility, the only fact that the NCDC can attest to is that exact duplicates of climatic records on file at this center have been provided to those that request such data. The standard Department of Commerce (DOC) or General certification of authenticity regularly provided to clients appropriately accomplishes this (see form examples).

In accordance with 28 U.S.C. 1733, properly authenticated copies or transcripts of records or publications of the National Oceanic and Atmospheric Administration (NOAA) shall be admitted in evidence as equal to the originals thereof. Although this statute only applies to Federal Courts, many, if not all, States have similar provisions.

The DOC gold seal with blue ribbon certification of record authenticity has for many years been provided to thousands of clients who successfully entered documents as court exhibits in legal proceedings. This DOC certification has been developed by the NOAA Office of General Counsel, in conjunction with the NCDC

and with the approval of each state's attorney general, to be universally acceptable to all U. S. federal and state courts. Only in extremely rare instances has this DOC certification been unacceptable to some courts, and then, there were only minor
points of contention. In order to maintain an expeditious flow of data to requesters, certain individuals at the NCDC have been designated by DOC/NOAA to sign these certifications in lieu of the original signatories. These "for" signatures have also been accepted without qualification by federal and state courts.

The DOC certification, or any affidavit, provided by the NCDC in no way authenticates or guarantees the data values contained in the records provided. The NCDC cannot attest to data
accuracy since it has no direct knowledge of observer expertise, instrument reliability, or the conditions under which the data were recorded. The DOC certification merely facilitates the admittance of the documents it is attached to as evidence in judicial proceedings. NOAA regulations prohibit employees of the NCDC from providing services to the public that would be in competition with private commercial enterprises. Prohibited services include consultation, interpretation, evaluation or any other services that are commonly

provided by certified consulting meteorologists or other enterprises in the private sector.

As data providers, NCDC employees can only advise customers of the availability and contents of data archived at the NCDC. Because of this prohibition, "Expert Witness" testimony cannot be provided by NCDC personnel. The NCDC cannot certify data obtained from any source other than directly from its archives, even if that data had previously been provided from those archives and returned to the NCDC for certification.

The DOC certification (Form CD-64) is a two part certificate (certification and authentication) under the Seal of the United States Department of Commerce with blue ribbon. This is the highest form of certification offered by the NCDC and can be attached to a total of up to 40 pages of data. This type of certification is signed by both the NCDC Records Custodian and the NCDC Director as the Certifying Officer, or their designated representatives.

The General certification consists of an attached certification statement signed by a Certifying Officer and is acceptable in most courts. The statement is affixed with a staple to the assembled documents, up to 70 pages. On rare occasions, customers request certifications with wording that differs from that of NCDC's standard certification. This service is available but is considerably more expensive and it increases the time required to respond to the request to several weeks. The specific wording must be edited to correspond to NCDC's guidelines and restrictions, returned to the requestor for review, forwarded to the NOAA Office of General Counsel in Silver Spring, MD, for approval, and returned to NCDC for signatures. One (1) certification can be attached to up to 40 pages of data.

While this certification information is designed to respond to the most frequently asked questions concerning weather records, it is recognized that it is not all inclusive. Special circumstances may occur that require specific answers to questions, in which case the NCDC should be contacted for further information or clarification.

Official weather observations are taken, either manually or automatically, at several thousand sites throughout the United States. The locations of these sites in any state may be obtained from any National Weather Service (NWS) Office in that state or from the National Climatic Data Center's web page at http://www.ncdc.noaa.gov/ol/climate/stationlocator.html .

Most stations record a daily precipitation measurement. At many stations, the daily highest and lowest temperatures are also recorded. Some stations have automatic precipitation gauges which provide a continuous record of rainfall. At approximately 700 NWS Offices much more detailed data are recorded at hourly intervals, such as wind, temperature, humidity, type of weather, atmospheric pressure, visibility, and clouds. Similar detailed data are recorded at some airports by the Federal Aviation Administration, airlines, the military, and contract weather observers.

The NCDC files forecasts and warnings (beginning July 1983), upper air data, accounts of major storms, radar images, satellite pictures, weather maps (both surface and constant pressure), observations from

ships, buoys, and aircraft, limited solar radiation data, a variety of summaries of weather conditions, and climatological publications from foreign countries. Much of the information for the U. S. is published in the Climatological Data and Hourly Precipitation Data for individual States/Areas and in Local Climatological Data for individual stations. A brief description of the most requested NCDC data sets is provided as Appendix A at the end of this section.

The original, filmed, and digital records of data from all types of stations are on file, and may be ordered from the NCDC. Information regarding specific data, services, and costs can be obtained by contacting the NCDC by one of the methods given inside the front cover of this brochure. Meteorologists or Meteorological Technicians are currently available by telephone for assistance from 8:00 a.m. to 6:00 p.m. Eastern time Monday-Friday. The Department of Commerce, including the NCDC, is required by Public Law 100.685 (Title IV, Section 409) to assess fees, based on fair market value, to recover all monies expended by the U. S. Government to provide the data to requesters.

Copies of materials will be appropriately certified upon specific request. If certified copies or additional certification of publications that carry a printed certification are needed, the type of certification desired should be specified in the original request. Uncertified copies cannot be returned at a later date for certification. A new order involving costs for the requested certified data is required. Orders for certified data can be placed on-line but current restrictions prevent electronic certification for on-line delivery. On-line orders for certified data are processed and certified offline. Delivery is usually within 5 - 7 business days by the U.S. Postal Service.

Regulations generally preclude NOAA personnel from rendering opinions in private litigation. It is therefore suggested that attorneys obtain the services of a private consulting meteorologist (PCM) who can give an expert opinion.

Although the sunrise/sunset times do appear in some surface observation data and monthly summaries available from the NCDC. The Department of the Navy, U. S. Naval Observatory, Massachusetts Ave & 24th Street, Washington, DC 20390 has a web site where astronomical data can be found (http://tycho.usno.navy.mil.) There are many private sector sources that can be contacted for data concerning the rising and setting of the sun and moon, twilight, moon phases, etc.

MOST UTILIZED OF THE NCDC'S PRODUCTS

Surface Weather Observations
These ground level observations are taken at specific, mostly airport, sites primarily for, but not limited to, use in the aviation field. Meteorological conditions are observed at least hourly and present sky condition, visibility, weather and obstructions to vision, pressure, temperature, dew point, wind direction/speed, and pertinent remarks. Other elements, such as relative humidity, are available for most major airport sites. The contents and format of these observations may vary depending on the site type and method of observation (manual, automated, etc.) Observations at some sites may be for less than a twenty-four hour period. Daily forms are available for several hundred sites.

Climatological (COOP) Observations
These are daily observations obtained from a volunteer network of trained observers. In addition to daily precipitation, most sites record the temperature at observation time plus the maximum and minimum temperature for the full twenty-four hour period proceeding the time of observation. Some cooperative observations also record daily river stages, evaporation, soil temperatures and wind direction/speed. Monthly forms are available for about eight thousand active sites.

Unedited Local Climatological Data
Summarizes surface weather observation and daily data for the entire month. The data are an unedited version of the first page of the Local Climatological Data.

Local Climatological Data
Monthly summary of temperature, relative humidity, precipitation, cloudiness, wind direction/speed, and degree days; also contains 3-hourly weather observations and an hourly summary of precipitation for most sites. Annual publications contain a summary of the past calendar year as well as historical averages and extremes. Monthly/annual publications are available for about 300 active sites.

Climatological Data
Monthly edition contains daily maximum and minimum temperatures and precipitation. Some sites provide daily snowfall, snow depth, evaporation and soil temperature data. Degree day data are contained in each monthly publication with the seasonal (July-June) heating degree day and snow data contained in the July edition. The annual edition contains monthly and annual averages of temperature, precipitation, temperature extremes, freeze data, soil temperatures, evaporation, and cooling degree days. Monthly/annual publications are available for each state or grouping of states.

Hourly Precipitation Data
Contains hourly precipitation amounts from recording rain gages. Maximum precipitation is presented for nine (9) time periods from 15 minutes to 24 hours for selected stations. Monthly publications are available for each state or grouping of states.

Storm Data
A chronological listing, by states, of occurrences of storms and unusual weather phenomena. Information on storm paths, deaths, injuries, and property damage are presented. An "Outstanding storms of the month" section highlights severe weather events with photographs, illustrations, and narratives. An annual tornado, lightning, flash floods, and tropical cyclone summary usually appears in the December edition, but may appear in a later issue. Monthly publication includes all states.

Monthly Climatic Data for the World
Contains monthly means of temperature, pressure, precipitation, vapor pressure, and sunshine for about 2,000 surface data collection sites worldwide. Monthly mean upper air temperatures, dew point depressions, and wind velocities are presented for about 500 worldwide observing sites. Monthly publication.

NATIONAL CLIMATIC DATA CENTER
151 PATTON AVENUE ROOM 120
ASHEVILLE, NC 28801-5001
TELEPHONE : (828) 271-4800
TD for DEAF : (828) 271-4010
FACSIMILE : (828) 271-4876
E-MAIL : ncdc.info@noaa.gov
WEBSITE : http://www.ncdc.noaa.gov
"A national resource for climatic information"
(Courtesy NCDC website and NWS)

Notwithstanding, there are other means of data retrieval which can yield better results due to the obscurity of the location in relation to distant reporting station(s). This could be achieved by visiting local neighbors who may keep daily records of local weather phenomena. Some homeowners are so weather savvy that they have installed a weather station which records weather directly to their computer. Check with local farmers, agricultural extension offices and Department of Transportation offices. Some educational facilities such as public schools often have some form of weather recording devices for educational purposes. Marine environs almost always have a wind report for boaters. This is where you'll find information which could support the distant reporting stations or show how the local topography can greatly influence local tertiary effects.

Although they may not be "official", it does help in obtaining the solution. If all else fails and you have limited data to work from, you may have to utilize your forecasting and analysis skills and use upper air charts to analyze the surface features. This is one of the more difficult ways to obtain data for court room use, but if your analysis is proven accurate through scientific reasoning and it's theoretically and operationally sound, then you have a good chance of proving your findings. Such was the case in the house burning described in the previous chapter. It was this technique where I had to improvise using microanalysis techniques to determine what the wind pattern was at the surface. With the growth of remote sensing, this is becoming less of a problem in this day and age.

Video: After I spent over 40+ hours on a case, pulling my hair out trying to justify why the weather occurred at a specific time, I was presented with evidence that was submitted an hour prior to the courts. The next door neighbor had recorded this big house fire which showed the glowing embers cross the two-track from the west! Where was this video 2 weeks ago!? Needless to say, I still got paid and the video was visual proof that my theory was right and further substantiated my testimony. Not all cases are like this but it was nice to have that video. So, when you investigate the location, ask around. See if anyone has a record either by pictures or video.

Which brings me to another point, a picture is worth a thousand words right? Well, take a trip and check out the area. Look for visible signs where the weather could enhance or retarded the accident scene. These are signs that you can't see without visiting the site. Take pictures of the location in a 360 degree panoramic view. Check out topographical maps from the internet or other source. You're looking for anything that can directly impact the weather in that area. Example: A car spun out of control and was hit by another car. There were injuries and lawsuits were filed. The driver of the first car went down a hill (15% downgrade) and over corrected due to the icy/snowy conditions. Since they were at fault, the driver of the second car who was not at fault sued the driver of the first citing negligence.

To make matters worse, the passenger of the first car was also pressing charges against her own friend! Police cited the first car for driving too fast for conditions.

When I concluded my inspection of the area, I found a tree line to the north which protected/shielded the winds, which were gusting to 30 mph from the North-west that morning. As the driver came over the hill, she encountered strong winds from the north-west because she left the protection of the tree line and experienced what we call windsheer. She over corrected due to the strength of the wind and that's how the accident started. It was snowy and the road had patchy ice according to police records. Case was dismissed. Putting evidence in terms the jury can understand can make a big difference. We all have over-corrected, even on dry roads.

Had I not investigated the area, I would have had no idea as to the terrain www.topomap.com, topography or the affect caused by the uneven terrain. I'll reiterate, always try to visit the site and if you can't, employ someone with video/camera capabilities.

A sample topographic map pulled from www.topomap.com

In search for the truth, we often employ other methods which could alter our own simplified findings. Confused? Don't be. Here's an example;

A slip & fall occurred on Christmas Eve morning hours (9 a.m.). She claimed that she lost her footing in 8" of fresh fallen snow and that there was glaze ice below the ice. The days leading up to the accident showed that it was warm, mid 40's F and rain had fallen the previous day. During the overnight hours, the passage of a cold front chilled the air with a temperature drop from 41F to 24F within 8 hours. The rain had stopped just prior to frontal passage (Fropa) and the winds increased in speeds I excess of 25 mph.

After .04" of rain and falling temperatures to 24 degrees, our basic review would indicate that the moisture froze and the onset of snow would indeed support the victim's claim that she slipped and fell. Besides the overabundance of medical claims, there was something else that didn't seem right. I reviewed the accident scene (3 years after the date of the accident) and took pictures. Reviewed them and compared notes to the weather observations and re-created the weather for that day. Oddly enough (and what are the chances of this happening) I was within 2 days from the same day of the accident and we had warm temperatures as rain was forecast. Only this rainfall was expected to exceed ½" within 12 hours, followed by cooling temperatures and, of course, snow. What better way to test my theory than to see it first hand.

I recorded the weather and took pictures before, during and after the rain as well as the onset of cooling temperatures and snow fall. One thing eluded me and that was the drainage in the parking lot. With over ½" of rainfall, the pictures showed that, over time, the asphalt was damp to the touch, not wet. I researched the criteria on asphalt paving construction rules for the state and measured the parking lot and found that the lot had exceeded the minimum requirements for drainage by an additional 42%. With sufficient drainage and pictures to show for it, I was able to conclude that there was sufficient drainage before the onset of snow. This means that there was no "glaze" of ice as the victim indicated in her deposition.

One additional note: Having researched the asphalt cooling thermodynamic laws, I came across a Pave-Cool computer program which shows the cooling rates of asphalt based on air temperature and sub-surface temperature. By inputting the variables I had observed, the program calculated the cooling rate of the parking lot from 42F degrees to 32F degrees to be 4.5 hours. Whereas the air temperature drop from 42F to 32F was only 1.2 hours. This finding further supported the negligible risk for ice formation on the surface of the parking lot. One last comment on this Pave Cool program: The program was initially designed for asphalt cooling above 180F degrees. After I consulted with the engineer, (who also happened to be the computer programmer), he was able to modify the program to suit my specific needs in this case (sub-freezing). Think outside the box!

It goes without saying that if you search for the truth; you often have to rely on your abilities to acquire information outside of your immediate profession and scope of knowledge. In researching, you'll come across information that could have helped a previous or current case if circumstances were different. Catalog and save anything you find that will assist you in upcoming cases. It will save you time. I have all my research saved in folders (Marine, Auto, Aviation, Construction, Human… and so on). It'll be much easier to find and confer upon in the future.

QUESTIONS AND DISCUSSIONS

1. When working on a case, what source of data would you use in a known data sparse area to help in your investigation? What would you exclude?

2. While you may not be an expert in the field of asphalt cooling, would you refer to this program or the dynamic laws of cooling in your assumptions of how ice may have formed (or melted)? Explain why or why you wouldn't use this method in a class discussion.

3. If you are unfamiliar with another area of expertise, would you offer information to the attorney for consideration or just keep your opinions to yourself and provide what was requested? Explain your answer in a class room discussion.

4. If you were asked to provide your expertise on a case that is in another state or across the nation (or world), would you make an attempt to find the answer or would you refer the case to another forensic colleague in that area or who has greater knowledge in that specific scientific field?

Chapter 3

Legal Cases, Terms and Concepts

Due to the visibility of the Broadcast Meteorologist within the community, there's a really good chance they will get a call or two on weather information which could eventually be used in discovery or even used in court. To relieve yourself of any undo hardships, it would be best to review your station policy or handbook, let the Station Manager and/or News Director know of your work/cases and to insure that there would be no conflict of interest. They would be the deciding party on whether you take the case as an expert witness or not. It wouldn't be wise to be an expert witness in a case against one of your stations' biggest advertisers. That would be just one example of a conflict of interest.

When you are called to be an expert witness, you'll have to have your CV in hand or already on file with your law firm you'll represent. You are there as an individual on your own knowledge and skill of your profession lending your expertise in a case that could decide the outcome. It may come out in a deposition where you work, but let them know up front that you are not representing the station in any way. If you wish to make it more formal, seek the advice of an attorney on what would be the best way to set up your business.

Every forensic meteorologist should have a working knowledge in the legal aspects of the court room and the process of trials from start to finish. For lawyers who are interested in knowing some of the internal workings of meteorology please reference (For Lawyers Only) at the end of this chapter.

Policy Statement on the Weather Service – Private Sector Roles (1991)

"The issue [of expert testimony for weather-related private litigation] is addressed in detail in Federal regulations (15 CFR parts 15a and 909.4) which state that NOAA employees will not provide such testimony and generally anticipate that the private sector will. However, exceptions exist where NOAA and the NWS could provide expert testimony, for example, in Government-related cases. This, of course, in no way precludes the private weather industry's recognized role to provide expert testimony in both civil and Government litigation"

Processes and Concepts in Civil and Criminal Litigation
(*Randall B. Christison Attorney at Law as presented at the American Meteorological Society Conference of Forensic Meteorology)
The Expansion of the Daubert Expert Evidence Doctrine Continues.

Recent Decisions in Federal and Louisiana Courts

Few areas of law depend to a greater degree on expert testimony and evidence than litigation involving environmental law and toxic torts. Verdicts in such cases often turn on the credibility that the judge or jury affords a particular expert witness and his opinion. Thus, it is usually critical that a litigant in such a case have his own expert's opinion accepted by the court under the appropriate evidentiary standards, while at the same time arguing that the opposition's conflicting expert opinion does not meet the threshold of the standard for admissibility.

The importance of this evidentiary issue is amply reflected in the fact that, in the last six years, the United States Supreme Court has granted certiorari for appellate review of three cases that each deal with the standard of admissibility for expert testimony. To appreciate the significance of the high court's actions, it must be understood that the Court accepts relatively few of the many cases for which review is sought, choosing only those it considers of suitable import based on constitutional and policy concerns. Thus, the Court's trio of visitations in such a short time span to this evidentiary issue of the admissibility of expert opinion is remarkable.

Until 1993, the prevailing expert evidentiary standard was taken from the longstanding decision in Frye v. United States, 293 F. 1013 (D.C. Cir. 1923), which required that a scientific theory only be generally accepted in the relevant scientific community to serve as a basis for expert opinion. But with the adoption of the modern Federal Rules of Evidence in 1975, the Supreme Court stated that the focus of expert opinion was statutorily altered, warranting a different standard to test admissibility. In the case of Daubert v. Merrell Dow Pharmaceuticals,

Inc., 509 U.S. 579 (1993), the Court stated that federal courts should decide the issue of the admissibility of expert scientific opinion by the application of five factors:

(1) Whether the expert's theory has been or can be tested for validity;
(2) Whether the theory has been subject to peer review and publication;
(3) The known or potential rate of error when the theory is applied;
(4) The existence and maintenance of standards of control; and
(5) The degree to which the theory is accepted in the scientific community.

The federal trial courts are to apply these factors as a matter of a "gatekeeping" role to assure that expert opinion be both relevant and reliable to assist the trial court or jury in rendering a decision. While the Court noted that the new rules of evidence are premised on a more liberal evidentiary thrust, these factors have served in practice under Daubert and the cases which have followed it to limit expert opinion to valid scientific theories and to weed out junk science.

The federal trial courts are to apply these factors as a matter of a "gatekeeping" role to assure that expert opinion be both relevant and reliable to assist the trial court or jury in rendering a decision. While the Court noted that the new rules of evidence are premised on a more liberal evidentiary thrust, these factors have served in practice under Daubert and the cases which have followed it to limit expert opinion to valid scientific theories and to weed out junk science.

After the Daubert decision, the Court considered as a separate issue in the case of General Electric Company v. Joiner, 522 U.S. 136 (1997), whether a different standard should apply in allowing expert opinion into the record as opposed to excluding admissibility. In that case, the lower circuit appeal court ruled that, while a trial court's decision to admit expert opinion is to be reviewed on the "abuse of discretion" standard, a court's exclusion of such evidence should be weighed against a "particularly stringent standard of review."

However, the Supreme Court again emphasized that the trial court as "gatekeeper" is required to make the fundamental determination as to whether there is an adequate nexus between the data presented and the opinion posed, or instead is a connection borne only of the "ipse dixit" (he himself said it) of the expert's baseless assumptions. Whether the trial court admits or excludes the evidence in this role, the Court held that the decision is subject solely to the "abuse of discretion" standard as a matter of review.

The last case in the "Daubert trilogy" is the Court's recent decision in Kumho Tire Co. v. Carmichael, 1999 U.S. LEXIS 2189 (March 23, 1999). In that case, the issue was whether the Daubert factors are to be applied only in cases where expert opinion deals with purely scientific principles or, instead, are equally applicable in other areas of technical expertise. The expert's testimony in Kumho concerned a general theory as to the reason an automobile tire had failed, resulting in a severe accident. Using the Daubert analysis, the trial court excluded this form of engineering testimony as unreliable, but was reversed by the lower appellate court. The Supreme Court reinstated the trial court's determination, stating that the Daubert factors were equally applicable to engineering testimony - and to all forms of expert testimony - to the degree that each factor was relevant in considering the particular expert opinion.

Since the Daubert decision, the United States Fifth Circuit Court of Appeal, which reviews cases arising in Louisiana's federal district courts, has considered a number of cases in which it has applied the Daubert factors. Two of the most recent decisions are Moore v. Ashland Chemical, Inc., 151 F.3d 269 (5th Cir. 1998) and Black v. Food Lion, Inc., 171 F.3d 308 (5th Cir. 1999).

In the Moore case, a truck driver claimed to have suffered a permanent lung dysfunction due to short-term exposure to a toluene-based chemical spill. One of his expert physicians was to testify that the specific lung illness was a condition termed "reactive airways dysfunction syndrome." Though the physician expert was recognized as a respected and distinguished practitioner in the field of pulmonary medicine, he could not show a credible scientific link to toluene exposure and the diagnosed disease.

While the Fifth Circuit Court panel of three judges reversed the trial court's exclusion of the evidence, the appeal court granted rehearing to consider the matter en banc. Relying heavily on the Joiner case, the court then held that the trial court had not abused its discretion in finding that the expert's opinion was unreliable, as based either on conjecture or inadequate evidence, in posing that the plaintiff's specific illness was linked to a short-term exposure to toluene. Similarly, in the Black case, the plaintiff claimed that a slip and fall incident in defendant's store caused trauma resulting in the disease "fibromyalgia," characterized by generalized pain, poor sleep, loss of concentration, and chronic fatigue. As the Kuhmo Tire case had been decided by the Supreme Court upon hearing of this matter, the appeal court first noted that under that case the Daubert factors were unquestionably relevant to the admissibility of the expert physician's medical diagnosis. The magistrate judge at trial of a defense motion to exclude the diagnosis allowed the expert physician to offer her conclusion into evidence, but did not specifically apply the Daubert factors in more than a generalized manner. In reversing the magistrate's decision, the Fifth Circuit noted that the magistrate had not applied the factors with specificity to test the expert's physician's methodology of determining that the plaintiff's trauma could have caused the diagnosed condition. Though the physician attempted to rule out other causes for the disease, relevant medical literature indicated no causative link between trauma and contraction of plaintiff's condition. In the court's opinion, had the magistrate conducted a proper analysis under Daubert and Khumo Tire, "the utter lack of any medical reliability of [the physician expert's] opinion would have been quickly exposed."

Lastly, and most significant in litigation in the state's courts, the Louisiana Supreme Court, in State v. Foret, 628 So. 2d 1116 (La. 1993), adopted the Daubert standard for testing expert testimony nearly immediately after it was articulated. Noting that Sections 702 of both the federal and the Louisiana evidence codes pertaining to the admissibility of expert opinion are identical, the state supreme court stated that it should follow Daubert on the same reasons stated therein. In Foret, the criminal defendant had been convicted of sexual abuse of a minor, based in part on the expert testimony of a child psychologist. The psychologist used an interview technique, developed for diagnosing the effects of child abuse to prescribe appropriate treatment, as a basis for an opinion that the victim's interview responses positively indicated she had been abused. But, the court cited to literature that stated the technique was unreliable in delineating actual abuse from fantasy or fraud, and was accepted as valid only for its intended purpose. Based on the Daubert factors, while the court did not exclude the testimony of its own accord, it remanded the case for a new trial with the admonition that the trial courts exercise its "gatekeeping" authority to closely evaluate any such testimony offered at retrial.

Daubert has been cited in reported Louisiana state court cases dozens of times. The most recent citation is in State v. George, 1999 LEXIS 447 (La. Feb. 26 , 1999), where the court sided with the trial court's exercise of the "gatekeeping" function to exclude evidence of defendant's alleged "limbic psychotic trigger reaction," asserted as a justification for violent conduct.

As seen in the foregoing cases, the admissibility or exclusion of expert opinion after Daubert can be decisive in an environmental or toxic tort case. Of significance to Louisiana businesses, a bill currently is being considered in the legislature (HB 247) which would impose a "loser pays" rule as to legal fees and costs against plaintiffs who bring frivolous lawsuits. The above decisions indicate that a challenge to the expert basis of a case under the Daubert factors can be key in exposing baseless litigation. Should the legislature pass this bill, the "Daubert challenge" may well be one method used to test the relative merits of a matter alleged frivolous by the defendant.
Courtesy: M. G. Durand of Onebane Law Firm

Often there are times when black ice becomes an issue in a defense. "Sudden Emergency" as lawyers would describe it.

Defendant's admitted knowledge of worsening weather conditions and presence of road hazards does not justify charging the jury with an "emergency doctrine" instruction. What follows is a case from the New York State Court of Appeals decision in the Caristo v. Sanzone matter.

Greg MacMaster   Environmental Forensics

Cases involving "Sudden Emergency" Defense

New York Court Opinions
Antoinette Caristo et al.,
Appellants,
v.
Augustine Sanzone et al.,
Respondents.
2001 NY Int. 37
April 3, 2001

This opinion is uncorrected and subject to revision before publication in the New York Reports.

Arnold E. DiJoseph III, for appellants.

Alan M. McLaughlin, for respondents. GRAFFEO, J.:
The issue in this motor vehicle accident negligence case is whether the trial court erred in charging the jury on the emergency doctrine. Under the facts presented, we conclude that defendants were not entitled to this instruction.

At approximately 9:00 A.M. on the morning of the accident, defendant Augustine Sanzone was driving a vehicle owned by his wife, defendant Patricia Cinquemani, on Foster Road in Staten Island. At the same time, plaintiff Antoinette Caristo was operating her automobile on Woodrow Road. Foster Road terminated at a "T" intersection with Woodrow Road, and a stop sign controlled the flow of traffic from Foster Road onto Woodrow Road.

At trial, Sanzone testified that the weather conditions at 7:00 A.M. that day consisted of snow, rain and freezing rain. This mixed precipitation was unchanged when he and his family left their home at approximately 8:30 A.M. By the time he drove to Foster Road, the weather had worsened. He described the conditions as "more like frozen rain and hail at the time." The temperature that morning was established, by stipulation of the parties, at 22 degrees Fahrenheit. After cresting an incline on Foster Road, Sanzone proceeded downhill toward the Woodrow Road intersection, traveling at 20 to 25 miles per hour. At this juncture, his vehicle began to slide and he noticed there was "a sheet of ice" on the hill. Despite Sanzone's effort to pump the brakes, the vehicle slid 175 to 200 feet, past the stop sign and into Woodrow Road.

As plaintiff approached the intersection at 15 to 20 miles per hour and saw defendants' vehicle, she attempted to swerve to avoid a collision, but was unsuccessful. Both Cinquemani and the police officer who responded to the scene of the accident confirmed the icy conditions on Foster Road. Neither plaintiff nor Sanzone experienced difficulty controlling their vehicles prior to this incident. Over plaintiff's objection, the trial court charged the jury on the emergency doctrine. The jury returned a verdict in favor of defendants and the trial court entered a judgment dismissing plaintiff's complaint. The Appellate Division affirmed the judgment, with two Justices dissenting (274 2 406). Plaintiff now appeals as a matter of right.

More than a century ago, this Court first considered the reasonableness of an actor's conduct when confronted with a sudden emergency situation (see, Wynn v C.P., N.&E. R.R.R. Co., 133 NY 575). Since then, we have articulated and applied the common-law emergency doctrine which "recognizes that when an actor is faced with a sudden and unexpected circumstance which leaves little or no time for thought, deliberation or consideration, or causes the actor to be reasonably so disturbed that the actor must make a speedy decision without weighing alternative courses of conduct, the actor may not be negligent if the actions taken are reasonable and prudent in the emergency context" (Rivera v New York City Tr. Auth., , 77 NY2d 322, 327), provided the actor has not created the emergency.

The rationale for this doctrine -- the need to instruct a jury that it may consider the reasonableness of a party's conduct in light of the unexpected emergency confronting that person -- has been somewhat eroded by the evolution from contributory negligence to comparative negligence. With the advent of the ability of juries to allocate fault and apportion damages, the viability of the doctrine has been questioned by some jurisdictions, with a few states going so far as to abolish it (see generally, Modern Status Of Sudden Emergency Doctrine, Ann. 10 ALR5th 680).

In New York, in addition to the elements of the charge, we have defined the role of the Trial Judge in assessing the propriety of an emergency charge request. We require the Judge to make the threshold determination that there is some reasonable view of the evidence supporting the occurrence of a "qualifying emergency" (Rivera v New York City Tr. Auth., supra, 77 NY2d, at 327). Only then is a jury instructed to consider whether a defendant was faced with a sudden and unforeseen emergency not of the actor's own making and, if so, whether defendant's response to the situation was that of a reasonably prudent person (see, PJI 2:14).

The emergency instruction is, therefore, properly charged where the evidence supports a finding that the party requesting the charge was confronted by "a sudden and unexpected circumstance which leaves little or no time for thought, deliberation or consideration" (Rivera v New York City Tr. Auth., supra, 77 NY2d, at 327; Kuci v Manhattan & Bronx Surface Tr. Operating Auth., , 88 NY2d 923, 924; see also, Restatement [Second] Torts § 296). Here, even considering the evidence in a light most favorable to defendant (see, Kuci v Manhattan & Bronx Surface Tr. Operating Auth., supra, 88 NY2d, at 924), we hold as a matter of law that there was no qualifying event which justified issuance of the emergency instruction. Given Sanzone's admitted knowledge of the worsening weather conditions, the presence of ice on the hill cannot be deemed a sudden and unexpected emergency.

Although Sanzone did not encounter patches of ice on the roadways before losing control of his vehicle, at the time of the accident the temperature was well below freezing and it had been snowing, raining and hailing for at least two hours. As such, there was no reasonable view of the evidence that would lead to the conclusion that the ice and slippery road conditions on the Foster Road slope were sudden and unforeseen. Defendants were not, therefore, entitled to an emergency instruction and the charge to the jury constituted reversible error under these circumstances.

The dissent contrasts our holding here with Ferrer v Harris (55 2 285), where we concluded defendant was entitled to an emergency doctrine charge. Ferrer is clearly distinguishable in that defendant was confronted by an unanticipated event when a four-year old child ran in front of his vehicle from between two parked cars. The qualifying emergency -- a child darting from a sidewalk into street traffic -- is simply not analogous to the presence of ice and slippery conditions following at least two hours of inclement weather with temperatures well below freezing. Accordingly, the order of the Appellate Division should be reversed, with costs, and a new trial ordered. Caristo v Sanzone No. 22 Rosenblatt, J. (dissenting):

By holding it was error to charge the emergency doctrine, the majority concludes as a matter of law that defendant expected to encounter a 175 to 200 foot sheet of ice while driving on an otherwise ice free road. In my view, it was for the jury, and not the court, to determine whether this sheet of ice was "unforeseen." I agree that the emergency doctrine should not be charged merely for the asking; nor should it be charged in every foul weather fender-bender case. But here the charge was justified. Accordingly, I dissent.

Because we are required to view the evidence most favorably toward the party requesting the emergency instruction (see, Kuci v Manhattan & Bronx Surface Tr. Operating Auth., , 88 NY2d 923, 924), we must accept the following: First, defendant drove onto a sheet of ice 175 to 200 feet long. Second, the character of the ice was such that, despite his efforts to stop the car, he slid the entire 175 to 200 foot distance -- a journey approximately two-thirds the length of a football field. Third, defendant was unable to see the sheet of ice until he was upon it. Fourth, defendant was driving only 20 miles per hour when he encountered the ice. Fifth, although the weather was bad (rain, snow, sleet and freezing temperatures), neither plaintiff nor defendant had encountered ice in the vicinity. In my view, this raises a question of fact as to whether defendant was confronted with an emergency.

It is settled law that an emergency doctrine charge must be given if, under some reasonable view of the evidence, the party requesting it was confronted with a "qualifying emergency" at the time of the alleged tortious conduct (Rivera v New York City Transit Auth., , 77 NY2d 322, 327). A "qualifying emergency" is a "sudden and unforeseen occurrence not of the actor's own making" (Rivera, supra, 77 NY2d, at 327 [citing Ferrer v Harris, , 55 NY2d 285]).

Given these facts, the jury could reasonably have concluded -- as they obviously did -- that encountering this enormous sheet of ice was "a sudden and unforeseen occurrence." Over the course of a lifetime, few winter drivers will ever encounter an ice hazard of that magnitude -- let alone when the roads are otherwise free of ice.

Accordingly, the trial court properly gave the charge. In doing so, it did not conclude that an emergency existed. It merely ruled that, on the evidence presented, the existence of an emergency was debatable and allowed the jury to resolve the point.[1] Charging the emergency doctrine simply reminds the jury that it must consider the reasonableness of a party's actions in light of the existing circumstances (see, Ferrer v Harris, supra, , 55 NY2d 285, 292 (citing Wagner v International Ry. Co., 232 NY 176, 182 [Cardozo, J.]).[2] As we noted in Rivera, giving the emergency doctrine instruction is "by no means" a directed verdict for the party requesting it (Rivera, supra, at 435) or even a declaration that there was an emergency. Rather, the jury still has the final say as to whether there actually was an emergency and, if so, whether the party reacted to it reasonably (cf., Rodriquez v New York State Thruway Auth., 82 AD2d 853, 854 [party's actions still unreasonable even though he was confronted with an emergency]).

In Ferrer v Harris (supra, , 55 NY2d 285, 290-291), defendant was driving his car 20 miles per hour on a block filled with children. Plaintiff, a young girl, darted into the street and was struck by defendant's car. Defendant testified that he saw children playing on the sidewalk before the accident. He even admitted seeing plaintiff step between two parked cars seconds before she darted (Ferrer v Harris, supra, 55 NY2d, at 290-291). The trial court refused the emergency doctrine charge. We held that the trial court's failure to charge it was reversible error because "it was more than conceivable that a jury could conclude that this defendant was faced with an emergency" (see, Ferrer v Harris, supra, at 292 [emphasis added]). More recently, in Rivera, we again reversed for failure to give the emergency doctrine charge. We concluded that "the jury could reasonably have concluded" that the accident was sudden and unexpected (see, Rivera, supra, 77 NY2d, at 327).

Indeed, we have consistently authorized if not required the charge so as to allow the jury to resolve whether an emergency occurred and whether the party's actions were reasonable in the face of it (see, Kuci v Manhattan & Bronx Surface Tr. Operating Auth., supra, , 88 NY2d

923, 924 [trial court committed reversible error by failing to charge emergency doctrine because driver testified that he "did not anticipate being suddenly cut off by this particular car"]; Mas v Two Bridges Assoc., , 75 NY2d 680, 686 [upholding trial court's emergency doctrine charge because "we think that on the evidence in this case the emergency was not dissipated * * * as a matter of law, and that the question was properly submitted to the jury"]; Amaro v City of New York, , 40 NY2d 30, 37 [emergency charge was properly given]).

Referenced from the New York Court of Appeals, CARISTO v. SANZONE, April 3, 2001. Presiding Judge: Judge Graffeo

I recognize that retention of the emergency doctrine has been under discussion nationally and that some jurisdictions have altered or abolished it.[3] The doctrine, however, is still a part of New York law. That being so, I submit that the trial court and the Appellate Division correctly applied it. The trial court and the Appellate Division majority concluded that the matter was at least arguable, which is to say that there was "some reasonable view of the evidence" that defendant was confronted with an emergency. The jury went even beyond that; as fact-finders they concluded that defendant was indeed confronted with an emergency and his actions were reasonable in light of it. Their verdict does not strike me as irrational.
Accordingly, I would affirm the order of the Appellate Division.
Order reversed, with costs, and a new trial ordered. Opinion by Judge Graffeo. Chief Judge Kaye and Judges Levine, Ciparick and Wesley concur. Judge Rosenblatt dissents and votes to affirm in an opinion in which Judge Smith concurs.

Litigation Proceedings

Listed below are types of Litigation Proceedings and categories that were presented by Sean Potter, CCM at the AMS Conference Workshop for Forensic Meteorologist, San Diego, CA 2005. Additions were added by the author. Because of the vast examples these categories could play, we have elected not to describe them here.

Civil
- § Personal Injury
- § Hourly Weather Data (Temperatures)
- § Snow and Ice accumulation
- § Visibility
- § High winds
- § Property Damage
- Wind data
- Rainfall records
- Thunderstorms
- Lightning
- Hail
- § Motor Vehicle Accidents
- Rainfall, snowfall
- Visibility (Restrictions to visibility)
- High winds (Wind Shear)
- Illumination data (Sun/Moon Data)
- Cloud cover
- § General and Commercial Aviation
- Aircraft icing
- Turbulence
- Wind Shear
- Thunderstorms
- Micro bursts
- Temperature (Warmer air requires longer take-off distance) (Overloaded)
- § Maritime
- Storms
- Winds
- Wave heights
- Visibility
- § Precipitation (Ice in winter, Hail in summer)

Construction Delays (Planning will be defined under Engineering – in a different chapter)
- § Thunderstorms/Lightning
- § Strong winds
- § Heavy rain/snow
- § Flooding
- § Temperature (Restricting certain construction procedures)

Air Pollution
- § Dispersion Models

- § Wind profile (Vertical and horizontal)
- § Stability
- § Toxic Corridor (Chemical dispersion/rate/amount/severity)
- § Criminal Investigations
- Insurance
  - o False claims filed
  - o Verification of damage associated with storms watches/warnings, black ice (sudden defense).
  - o Some cases may be investigated by Insurance companies without ever seeing the inside of a court-room.
- Environmental Impact Assessment

Legal Jargon and Definitions:

- § Parties: In civil cases, the Plaintiff is the person or entity that filed the lawsuit and the Defendant is the person or entity being sued.
- § Expert witness vs. Percipient Witness: Experts can rely upon hearsay information, and may express opinions based upon their training and expertise. Percipient witnesses can only testify to what they have seen, heard, etc.
- § Discovery: Expert Deposition. A process where the expert witness gives testimony, under oath, regarding their expert opinions and the reasons for the opinions.
- § Trial Participants:
- § Judge: The person who presides over the trial, and makes legal rulings on issues that arise during the case.
- § Marshal: The law enforcement person who maintains security.
- § Clerk: Is the person who handles exhibits, pleadings and calendaring in the court room.
- § Reporter: Is the person who prepares a shorthand record of all trial testimony.
- § Jury: The individuals approved by both Plaintiff and Defendant Attorneys who decide the case.
- § Voir Dire: The process by which judges and lawyers select a petit jury from among those eligible to serve, by questioning them to determine knowledge of the facts of the case and a willingness to decide the case only on the evidence presented in court. "Voir dire" is a phrase meaning "to speak the truth."

- § Hypothetical Question: A question that sets forth certain assumptions, and asks for an expert opinion based upon those assumptions.
- § Negligence: Doing something that a reasonable person would not do, or not doing something that a reasonable person would do.
- § Reasonable scientific probability: What is required for a scientific expert opinion?
- § Beyond a Reasonable Doubt: The burden of proof for the prosecution in a criminal case.
- § Preponderance of the Evidence: The burden of proof for the plaintiff in a civil case.
- § "Junk" science: Theories that are not supported by relevant scientific literature and practitioners.
- § Verdict: The decision made by the jury. This is later turned into a judgment.

Opinion Testimony & Scientific Evidence

(*Karen Holmes Balestreri, Pendleton & Potocki as presented at the AMS Conference of Forensic Meteorologist. Jan/2005)

California Evidence Code §801

If a witness is testifying as an expert, his testimony in the form of an opinion is limited to such an opinion as is:

(a) Related to a subject that is sufficiently beyond common experience that the opinion of an expert would assist the trier of fact; and

(b) Based on matter (including his special knowledge, skill, experience, training, and education) perceived by or personally known to the witness or made known to him at or before the hearing, whether or not admissible, that is of a type that reasonably may be relied upon by an expert in forming an opinion upon the subject to which his testimony relates, unless an expert is precluded by law from using such matter as a basis for his opinion. Leg.H. 1965 ch. 299, operative January 1, 1967.

B. Federal Rules of Evidence §702

If scientific, technical, or other specialized knowledge will assist the trier of fact to understand the evidence or to determine a fact in issue, a witness qualified as an expert by knowledge, skill, experience, training, or education, may testify thereto in the form of an opinion or otherwise, if (1) the testimony is based upon sufficient facts or data, (2) the testimony is the product of reliable principles and methods, and (3) the witness has applied the principles and methods reliably to the facts of the case.

I should mention this simply because it has helped me tremendously and that's having the advice of a lawyer who is looking out for your best interest. I'm not talking about the lawyer your representing as an expert witness; I'm talking about someone who is on the outside looking in. There are a number of legal benefit companies that offer such services but you usually need to be employed by a company in order to receive such benefits. One such company Pre-Paid Legal Services offers such a service to the individual. I only pay $25/month and I can call my lawyer on any issue. If I'm worried about a conflict of interest, what my liabilities are, risks involved… I have my bases covered. Plus, they're services go far beyond consulting which are included in that monthly fee. Contact http://www.WLSorg.com/ to find out more. Or you can contact Pre-Paid Legal directly at www.pplsi.com .

There are other companies out there and I'm not familiar with them as I have only used this particular service. Feel free to shop around.

Greg MacMaster   Environmental Forensics

Sample Expert Testimony Letter

Name
Title
Business
Street
City, State, Zip
Phone
E-mail

Date:

Lawyer
Address
City, State Zip

RE: Plaintiff v. Defendant

Dear Mr. Lawyer,

Per your instructions, I have reviewed the necessary meteorological records, supplemental information from local, state and governmental authorities & coop weather observers that surrounds the area (accident) and am submitting my conclusions for your review.

Overview:

You asked about the weather and surface conditions that led up to the date of the accident – in this case -a slip and fall in the parking lot, on December 24th, 2001 at a business called XX. You asked me to look into the possibility of ice forming prior to the snowfall during the early morning hours of December 24th, which could led to the cause of the accident.

Data:

The amount of official weather data from observation locations is limited due to the population of (region). Considering the close proximity of the business to the local airport, data from the ASOS (Automated Surface Observation System) was used. It would be inappropriate to use data from other regional ASOS stations as the micro-climate of (Region) can vary tremendously, especially when locations are close to a large body of water (in this case, XX Lake). I reviewed the observations from the airport as well as local Department of Transportation logs showing county snowfall as well as the snowplow log from the company hired to maintain said parking lot.

Discussion:

This where you would get "down and dirty" with your reasoning and theory. Using all data available and putting it in a logical, easy to understand format. No junk science!

I added one of the formulas I have used in past cases:

Using Stefan-Boltzmann Law: If object at temperature T, in this case, Asphalt, is surrounded by an environment (air) at temperature T0, the net heat flow is:
$H_{net} = esA [T^4 - T0^4]$

This is why the onset of snow would melt when making contact with asphalt.

The opposing expert testimony would only be correct for a lot with no drainage (standing water) and conditions where the atmosphere is in (a state of equilibrium) where no mixing (evaporation/sublimation) would occur. This (state of equilibrium) is theoretically, scientifically and operationally impossible as there's always some form of energy flux within the atmosphere, especially given the strength of the surface wind flow and drastic change in temperature (as indicated by surface observations) to induce sublimation.

Supporting Documentation:

Plow records from County (MDOT)
Snowplow billing record from XX
Weather records for XX Airport for December 23 and December 24th, 200X
Pictures showing conditions before rain, during and after rain. Date and times and current weather noted at bottom of pictures. (Note: Sufficient draining)

Conclusions:
Awaiting info from XXDOT about drainage info…..

I certify that the above information is true and accurate and that any estimations, interpolations or assumptions that have been made were done so with expert accuracy by a professional meteorologist. Additionally, I reserve the right to amend these conclusions made herein upon further discovery of meteorological data.

Sincerely,

Name
Consulting Meteorologist
XX Company

The suggested format is merely a sample as you will want to change or modify to suit your needs. But it's a good start if you have nothing to reference from. As your knowledge grows and you develop more cases, you'll develop a laundry list of additional comments, reference to manuals, operating procedures and map/graphics/pictures. Notice the last paragraph:

"I certify that the above information is true and accurate and that any estimations, interpolations or assumptions that have been made were done so with expert accuracy by a professional meteorologist. Additionally, I reserve the right to amend these conclusions made herein upon further discovery of meteorological data."

If you come across information that could sway your testimony, you have the right to make changes. Don't hold back information that could sway the case, you need to be honest, ethically and morally correct. Words like "Possible" or "Alleged" are not words that should be used in testimony. Be accurate and have sound theoretical, mathematical, scientific or operational reasoning backing your testimony. It's your reputation that's on the line.

What you need to get started.
(Parts presented by Jan Null, CCM, AMS
Conference Forensic Workshop,

San Diego, CA 2005)

Before the phone rings, you need to be ready to provide the necessary documents that will define your ability to work as a forensic meteorologist. Here's a list (basic) that you need to keep current:

- Resume
- CV (Circum Vitae)
- Fee Schedule
o Rate Structure
o Travel Expenses
o Billing

A resume' is needed to define your accomplishments based on years of experience. Don't overstate your qualifications! That's the quickest way for the opposing counsel to shoot holes in your testimony.

Jan Null, CCM says "A CV is not necessarily a shorter version. It just has a different focus and leaves out a "career goals" section. A good overview can be found at: http://www.brandeis.edu/hiatt/cv.html"

Usually listing your specific reports/papers that were peer reviewed. Manuscript(s) that was published in professional magazines and such. Speaking engagements and previous cases that you worked that defined a new law. Information that shows credibility in your current profession should be listed.

Fee Schedule varies depending on location. Check your other competitors or ask a lawyer in your area what if feels is adequate for the area. This is best calculated using a spreadsheet (Microsoft Excel) or a similar program and has it available to fax, e-mail/present to your requesting party.
Some expert witnesses are leaning towards a flat fee schedule.

Professional Rate: $xxx per hour
[3 hour minimum]

Greg MacMaster   Environmental Forensics

Terms:
- A 3-hour minimum is to be paid within 15 days of engagement as a retainer for services. This fee is non-refundable.
- Travel time is billed as professional time.
- One hour of billable time will be charged for any cancelled appointments, depositions and court dates if less than 48 hours notice is given.
- When the client requests services on weekends and holidays there is a 30 % surcharge.
- Data purchased from external sources and expenses are billed at cost.
- Automobile mileage will be billed at current IRS mileage rate.
- Air travel is billed at coach fares.
- Billing is every 60 days for all work and expense to date.
- Invoices are payable within 30 days of receipt.
- A finance charge of 1.5% per month will be added to invoices not paid within 60 days.

Billing should accurate with a complete list of your time and how it was divided. Don't pad or over bill. Keep an accurate record of your work. A good way to track your total time is to purchase a stop watch that can count up or down. Start the clock the second you leave your front door and stop it when you return.

Cold Calling: It wouldn't hurt if you made an effort to meet some of the lawyers/law firms in your area. Have business cards, a copy of your CV, Fee Schedule and ask what cases they may be working on at that time that may be questionable as to whether the environment played a role in the accident. Advertise in their quarterly bulletin. Numerous lawyers are unaware as to how the weather can affect a crime scene or accident. I'll touch on this later in this chapter.

So you've received your first call about weather information. Did you get all the pertinent information about the case? Having a preprinted contact sheet with the basics would help. Here's a list of the basics;

Firm Name
Name of Attorney
Firm Address

Greg MacMaster					Environmental Forensics

Phone number
Fax
E-mail
Case# (or File #)
Plaintiff v. Defendant involved:
Exact Time and Date of Incident
Location of Incident
Ask what pertinent information they are requesting.
Deadline

If you are called as an expert witness, you may simply get a call stating that you have been added on their case. Be sure to log that information as you may receive another call from the opposing law wishing to use you as their expert witness. That would be a conflict of interest. Log each call by "Case/file" number or "Plaintiff/Defendant" or "Jurisdiction Court".

With all the necessary information in hand, you now are on your way to your first of many cases in your career. Your list of resources will determine how long it takes for you to gather vital information to your case. Get to know anybody who has an interest in weather. Farmers who keep daily records, plow companies, harbors, anybody! This will come in handy when the time comes where you need to cross reference your weather data from a distant recording station. Try to use public data that you know has been calibrated or quality controlled. Some sources would be the US Forest Service, National Weather Service, local water districts and the data should be freely available (not to say it's free - but available to both parties).

I had attended a business expo in my area and met with a number of businesses as well as some law firms who were represented by local counsel. This is a great way to drum up business! One of the lawyers asked what I did and I told him I was a forensic meteorologist. He looked puzzled and asked what that was. After a quick summary and sample cases I worked, he was surprised and then said that his firm doesn't need the advice of a forensic meteorologist. After we exchanged business cards, I realized his downstate counterpart was the opposing counsel on a case I was called to testify on. It was an automobile accident involving snow, ice and wind shear. I asked him how that case was decided (knowing full well how it ended), he said that he wasn't sure but that it may have been dismissed (They lost their case). I told him I was the expert witness hired by the defendants' counsel. His facial expression was priceless! I shook his hand and bid him a good day.

As a forensic meteorologist, you need to have the foresight to see how the weather can play a factor and explain that to the counsel. Know your formulas, thermodynamic principles and be ready to learn parts of other professions.

The Process
Initial contact
Usually by letter, phone call or e-mail asking for information. This is also the time where you should log the contact and deter anyone else from hiring your services since they were the first to contact you. As this would be a conflict of interest.

Phone interview:
Usually to review the "meat" of the case and what they're looking to do. Face-to-face meeting to go over litigation and strategy:
Usually conducted in the lawyer's office or conference room depending on what part of the nation you're in. Review all documents pertaining to the case, previous depositions, testimony, review citations if available and formulate a plan of strategy.
Citations from various law enforcement departments can vary greatly. It's this variance that can often create holes in a testimony and allow the opposing or defense team to pursue legal action. The more weather information that can be recorded at the crash/crime scene, the margin of error is reduced.

Expert's research:
Forensics is: Fact based precision documentation. Examine the opposing expert witness testimony (if available). Conduct your own research, visit the site, gather data and start formulating a rough draft of how you want to prepare your thoughts for the report.

Preparation of expert report:
Due to discovery rules, find out of your attorney wants any written work, including correspondence by e-mail. A sample report is in this chapter which gives a basic outline how to proceed and the format that is generally accepted in a court of law.

Review of report with counsel, possible revision:
Once you have all the materials, copies of data, pictures available. Present this to your counsel for a review to see if it meets the criteria they are using in a defense case (or Plaintiff's case). Some reports may be no more than a few pages and I have had a few that were more than 150 pages. It all depends on the supporting documentation and evidence available.

Preparation for deposition:
After your report has been viewed by the opposing counsel, they may want to take your deposition. This is a process where the opposing lawyer (or more) will want to verify your credentials and try to bring discredit to your testimony. This is not the time to fall apart! Keep your cool and make sure you have analyzed your investigation from BOTH sides. If you're hired by the defense team, take time to review your report as if you were on the plaintiff's team. Look for mistakes, any documentation that you can't support through solid mathematical formulas, Empirical formulas, Operational analysis or weather fundamentals, shouldn't be in the report. Either find a solution or don't use it. No junk science!

Deposition process:
As stated above, this could take an hour, or some can go as long as 6 hours. It all depends on how complicated the case is and how many are named in the lawsuit(s). Keep your cool! There will be times when you go up against lawyers who have no knowledge in weather. Questions almost seem to be stupid in nature, but they don't understand the atmosphere like we do. I have asked that the stenographic "can we go off the record for a moment please?" so I can give a crash course in Meteorology 101. Showing how the weather works, (for instance, why is a prevailing westerly wind from the west) levels the playing field for everyone at the table and then you can continue on. Some depositions have been very formal with "yes" and "no" answers, others have been like a classroom environment where I was teaching more than I was answering.

You'll get all kinds of questions, some very direct and some extremely vague. Also expect them to question you 5 different ways

on the same topic. It's the job of the attorney who is deposing you to find holes in your testimony, so be prepared to be tested.
Be professional and courteous and well dressed and have all of your notes available. Leave nothing to chance.

Tips on testimony:
Direct answers only; don't volunteer unsolicited information from opposing counsel, consistency of professional opinions, etc.)
Review and editing of stenographic deposition record
"Woodshedding" (pre-trial preparation)

The chances of it going to trial are slim. But, just in case, you'll have plenty of time to review your report with counsel and review the depositions of all parties involved. By the time the case does go to trial, be sure to review them again as this process can take a long time and you could have very well provided a dozen or so cases.
Courtroom procedure and decorum
The trial may take anywhere from a day to week. Your lawyer will ask you to be available on certain days but won't be able to pin down an exact time because the trial is constantly in flux. So have your schedule open on those days. You'll most likely be asked to sit in another room or outside the courtroom proceedings until you are called to testify. When called, follow your lawyer's advice and directions.

Just like the deposition, come prepared with the materials you used to come to your conclusion. If you forget to bring a few things, no problem, you can ask them to jog your memory on some exhibits that have been admitted as evidence. Anything else you bring can be admitted as discovery evidence (even a woman's purse!). Your presentation style may vary, it could be as simply as a piece of paper to a power-point presentation. Whatever you need to support your testimony, have it available.

Be polite and professional and should your cross-examination from the opposing counsel get under your skin, don't let it bother you. That's their job! You may pick up on a question asked 5 different times in 5 different ways. The answer will be the same. They're there to trip you up. Be patient and calm and be a professional.
This may shock you but the way you present yourself and what you wear has a deciding factor if the jury is going to like you. Body

piercing, tattoos and the like are not going to fair well with the jury. You need to be professional, a suit for the man and a skirt below the knees for the women. For the man, do not wear a red tie as this shows a sign of aggression. Keep it a color that is consistent with the rest of your suit. For the women, wear your hair up. Again, look professional.

For Lawyers Only

Every case has its uniqueness, yet there are similarities in most winter accidents and that's ice, black ice (man made), packed snow or wind shear (at least in my cases anyway). There's one element in my investigation that has always assured a success in the end and that's scene evaluation. Either by pictures, (aerial or preferably ground), topography charts visual means. If your expert witness is local, this shouldn't be a problem. However, if you're seeking someone that works nationally and can't be at the scene, it would be a good idea to offer up some pictures, video or something that can give him/her a better idea of the scene of the accident. For without this, testimony may be incomplete or flawed.

Sudden Emergency, provided by Mother Nature or Man Made. That's right! Man Made. Here's how it works:

[Black ice is a thin formation of ice that occurs when water freezes on a road surface. Because it is so thin and clear, the dark underlying road surface easily shows through, lending to its common name. Black ice can naturally form at any temperature at or below freezing, but it is favored to form when the road surface is between 25F and 32F (-4C and 0C). This range was observed on February 10th in the late evening hours during snowfall.

A deicing compound, salt (Sodium Chloride), magnesium, etc. is applied in a concentration or mixture that will melt ice or snow on the highway at temperatures which may be much below the normal range of black ice formation. The applied compound starts melting the packed snow or ice on the highway. But, if the temperature does not rise, or worse, falls, the compound, diluted by the melting action is now below its working temperature and will start to freeze. The Department of Transportation will scrape/plow the roads and apply salt when accumulated snow is at 1 inch.

If left unattended and more concentrated deicing agent is not immediately applied to strengthen the dilution back to a point where it again melts, - "black ice" forms on the roadway. In the event the temperature falls and no precipitation occurs, (and in this case, the temperature dropped to the low 20's) the lower temperature may now be below the liquid point of the applied deicing agent, which would result in a thin layer of frozen or slippery 'deicing' compound that is now 'black ice'. If the humidity drops to a level where equilibrium with the saturated solution is reached, and the humidity continues to drop further for a longer period of time, (2 - 5 hours) [Note the drop in relative humidity prior to the accident] a portion of the calcium chloride solution on the road is converted into a solid form, (hexahydrate - $CaCl_2 - 6 H_2O$). This causes "ice to develop".

This can be a very dangerous condition because where there was packed snow or traces of ice and snow that was visible to the driver has now dissolved into a thin liquid film that refreezes and is not readily seen, and is unexpected at the low temperatures! This is more prevalent in and around intersections. The continuous snowfall with periodic wind gusts to 15 mph suggest that there was minor blowing snow over the road where the roads was not protected.]
(From the files of Greg MacMaster, Forensic Meteorologist)

More food for thought; Accidents on asphalt where the air temperature cooled below 32 degrees and visible moisture was present would give a perfect case for ice to occur. In my research for the cooling aspects of asphalt, I came a cross a computer program called PaveCool. The programmer had to modify it for temperatures below freezing, but it did the job and our defense team was happy with the court ruling. PaveCool shows the delay of cooling based on many factors, one being air temperature, thickness of the asphalt, grade and wind speed. In one case I found that the surface temperature of the asphalt cooled slower than the air, which delayed the onset of ice – by 3.4 hours. To access this program – go to: http://groups.yahoo.com/group/forensic_meteorology/ or contact the author at sales@forensicweatherman.com .

More tidbits of information: Drainage of parking lot or roadway, anything to show the drainage capability of the scene of the accident. Too many testimonial reports are fixed on a fact that isn't true in most cases. Christopher Columbus found out that the world wasn't flat and neither is a parking lot. State construction guidelines dictate the amount of slope (usually 1/100) or 1 foot of drop for 100 foot of run. Anything greater than that and you may not have standing water freezing!

In winter, salt/calcium chloride is often distributed on main roads to keep them from freezing. As vehicles drive in driveways, parking lots and other areas not usually treated, they'll usually carry a certain amount of salt/calcium chloride into the area. Although it's diluted, it can be enough to aid in a delayed onset of ice. Now I'm not trying to give a defense to all your cases, just broadening your scope of thinking in an altered environment. Weather isn't always black and white, but a gray scale. We have mathematical formulas, hypothetical reasons and operationally proven conclusions and if the right testimony is provided, there should be no margin for error and should be iron clad. The items I detailed above are more for the Defense counsel. We're in a litigious society and a good defense beats any offense.

I have reviewed testimony from a fellow professional and it would seem to reason that there would be a line of reasoning consistent with each other. Sometimes that's not the case. You may receive an expert witness testimony that describes exactly what you (the lawyer) is looking for – You need to be careful as you'll open yourself to a resounding display of opposing documents that could refute your original testimony and place doubt in your mind to his/her abilities.

Engineering Physics for Lawyers
 Much of the work of accident analysis and reconstruction involves the application of general physical laws to the particular situation at hand. These physical laws are expressed in terms of Newtonian mechanics. Newtonian mechanics is the base science of physics and engineering. Today the development and application of basic Newtonian mechanics is almost exclusively the province of engineers in that physicists conduct basic research only in "new physics" areas such as quantum mechanics, relativistic physics, etc.
KINEMATICS

The analysis of motion

Velocity. Velocity is the rate of change of the position of a body over time (velocity = distance/time). Velocity is a vector, which means it has both a magnitude and direction. Thus, 30 miles per hour north, or 20 meters per second along the X axis, are both velocities. Note that the more common term, Speed, is a velocity without a direction, so that 30 miles per hour or 20 meters per second are speeds, and not velocities, since no direction is specified.

Acceleration. Acceleration is the rate of change in velocity (acceleration = velocity/time). Thus, any time a body changes its rate of travel or its direction of travel, it is said to be accelerating. Acceleration is a vector requiring both magnitude and direction. Thus, for example, 32.2 ft/sec/sec., downward toward the center of the earth, is the acceleration vector of all bodies on the surface of the earth. If the support collapses, for example, if a person falls off a ladder, they will increase their velocity at a rate of 32.2 ft/sec. per second. At the end of one second of free fall, they will be traveling at a rate of 32.2 ft/sec. which is approximately 22 miles per hour. They will have fallen approximately 16.1 feet in this one-second interval. (D = 1/2 a t x t = 1/2 x 32.2 x 1 x 1 = 16.1).

NEWTON'S LAWS
First Law: "A body does not alter its state of motion without the influence of an external force." That is, there is no change in the velocity of a body (neither in magnitude nor in direction) unless some force acts on that body.
Second Law: "The net resultant force applied to the body is equal to the first time-derivative of the momentum function." Or roughly: Force = Mass x Acceleration. This relationship is not so much a natural law as a rule for assigning a magnitude to forces.

Third Law: "For every applied force there is an equal and oppositely directed reactive force." You push on the wall- the wall pushes back. The pusher and the pushed, the striker and the struck both experience forces of the same magnitude but of opposite direction.

DYNAMICS
Force. A force in physics or engineering is something close to what we call "forces" in everyday life. Any push, pull, twist, etc., involves

a force or forces. Forces are measured in physics according to their effects-- according to Newton's Second Law, Force = Mass x Acceleration, or F= MA. Note that "mass" is that property of physical objects that causes them to resist changes in motion. "Inertia" is another term for mass.

Weight. Weight is a special force which results from the fact that a mass is being acted on by a gravitational field. A body on the surface of the earth has a certain weight because it has mass which, when acted on by the earth's gravitational field, would cause the body to be accelerated if it lost its support. If it could fall, its "weight" would result in its being accelerated toward the center of the earth.
Torque. A torque is a twisting force. It is a force that tends to induce rotary motion rather than straight line motion. Torque is typically measured in lb-ft. Thus, if we pull on a one-foot long wrench with a force of 100 lbs., we exert 100 ft-lb of torque on the nut. The same torque is generated by 50 lbs. on a two-foot wrench, etc.
Friction. Friction is a special kind of force produced by two bodies that are in contact. If a book is at rest on a desk and we try to push it, our efforts are resisted by what is known as static friction. Once we get it moving, if we stop pushing it, it comes to rest almost immediately. The retarding or stopping force is known as dynamic friction.

Rolling Friction is the relatively low retarding force associated with the free rolling of objects, e.g. tires. Sliding friction is much greater than rolling friction and occurs whenever two objects in contact move with respect to one another without the benefit of any revolving elements. When a car is driven down the road without braking, it is being retarded by rolling friction (also by air drag, another type of force). When the wheels are locked, the vehicle is retarded by the sliding friction between the tires and on the road. In between these two cases, in situations with braking without wheel lock-up, the retarding forces are a complicated manifestation of the operation of the brakes, tires, and suspension system of the car. One measure of the sliding friction of cars is the coefficient of drag (Cd). This gives an indication of how hard it is to push the car along the road with its wheels locked. A typical Cd is 0.7.

In this case, a force equal to 0.7 x the weight of the car is required to keep it sliding along the road. With a Cd and skid mark of known lengths, it is possible to estimate a vehicle's velocity before the start of the slide. S(mph) = 5.5 x sqrt (Cd x length of skid)

Momentum. Momentum is the product of mass and velocity (momentum = mass x velocity). Momentum is a vector quantity as are velocity, acceleration and force. Momentum is conserved in impacts. That is, the sums of the various momentums of the bodies before the collision are the same as momentums after the collision. Thus, we can frequently compute the velocities of vehicles before a collision by knowing their speeds and directions of travel after the impact. Momentum is not the same as, and should not be confused with, energy.

Work.
When a force acts on a body, work is said to be done on that body. The quantity of work, W, is given by the formula: W = F x D, where "F" is the force that acts on the body and "D" is the distance through which it acts. Thus, a 100-lb. force acting for a distance of 10' (e.g., pushing an object against a resistance of 100# for 10') results in work = 1000 lb-ft being done on the object.

Energy.
Energy is the capacity to do work. In Newtonian physics, the energy of a body is computed in two ways: either by computing its kinetic energy: KE = 1/2 M x V x V where "M" = mass and "V" = velocity; or by computing its potential energy with respect to a system of forces capable of doing work on the body. For example, the potential energy of a body in the earth's gravitational field is PE = W x H, where "W" is the weight of the body and "H" is its height above some reference point.

Computations involving kinetic energy are tricky but informative. For example, if a moving body crashes into a solid, non-yielding wall at 20 miles per hour, the kinetic energy dissipated in the crash = 1/2 M x 20 x 20 = 1/2 M x 400 = M x 200; but if it crashes into the wall at 40 miles per hour, the energy dissipated in the collision is 1/2 M x 40 x 40 = 1/2 M x 1600 = M x 800. Thus, four times as much energy is involved in the second crash as in the first (M x 800/M x 200 = 4). The speed has doubled but the energy has quadrupled!

Power.
Power is the rate at which work is done. When work is done at the rate of 33,000 ft-lbs. per minute, one horsepower is produced. A "horsepower" is merely a conveniently sized unit for measuring power. Its connection with equine work capacity is tenuous. A human being can apparently work at a rate of about .35 horsepower for short periods of time, if all their skeletal muscles are being effectively used.

SPECIAL TERMS & EXAMPLES OF NEWTONIAN ANALYSIS

G's. The term "g" forces or "g" loads is a convenient descriptive tool for technical discussions involving accelerations due to impact forces.

One "g" is an acceleration equal to that generated by a free fall in the earth's gravitational field, i.e., 32.2 feet per second per second. Thus, a body acted on by a 0.5 g acceleration experiences a force equal to half its weight. This force acts in the direction of the acceleration. Frequently, during high speed collisions, accelerations in the range of 25 to 50 g's are generated. If a 150-lb. person is acted on by a 50-g retarding force during an accident, then a force of 7,500 lbs. acts on his body. Note that while a "g" value is really an acceleration, it is sometimes discussed as though it were a force. This is technically inaccurate but not harmful if it is clear what body the force acts on and if the mass of that body does not change during the application of the "g load."

Moment.
The word "moment" in physics is really another word for torque. Generally, when we have two or more torques acting on a body, the result is said to produce a "moment" or net torque, which acts to rotate the body in a direction determined by the combined torque vectors.

Pressure.
Pressure is force per unit area. Thus, if a 10-lb. weight has a contact area with another body of 1 square in., the pressure is 10 lb per square in. Normal atmospheric pressure at sea level is about 14.7 lb./sq.-in. (psi) This is the weight of a column of air 1 inch square extending up from sea level to the outer limit of the earth's atmosphere--roughly 80 miles high. A barometer reading of about 30 inches of mercury (30 in hg) is about one atmosphere pressure or 14.7 psi. Thus, a 30-inch

column of mercury weighs about as much as 80 miles of air since they both exert the same weight force per unit area.

Stress.
Stress, like pressure, is also a force per unit area. However, while pressure acts on bodies, stresse act in a body. If you pull on both ends of a steel bar of 1 in. cross-sectional area with a force of 100 lbs., You'll generate a tensile stress inside the bar of 100 psi . If I pull across the bar so as to try to split it in half, I generate a sheer stress. Note that a mild steel bar would be able to absorb tensile stresses in excess of 50,000 psi without breaking so that my efforts to pull it apart result in only trivial stresses in the bar.

ACCIDENT RECONSTRUCTION EXAMPLES

A 3000 lb. car crashed into the rear of a truck that weighs 30,000lbs. The road is asphalt and the coefficient of drag is 0.7, the car is going 10 miles per hour, the truck is at rest.
The momentum of the system before the collision is:
Momentum = Mass (car) x Velocity (car) + Mass (truck) x Velocity (truck)
= 3000 x 10 + 0 x 30,000
= 30000

Since the external forces acting on the system during the impact are minimal (the brakes of the truck are off) momentum is conserved. Thus, after the accident the following relation holds true:
Momentum before =Momentum after
30,000 = 3000 x V + 30,000 x V
V = 30,000/33,000 = 0.9 mph

where "V" is now the post impact velocity of the vehicles. (We assume here that the truck and the car are moving with the same velocity after impact)

Thus, the car loses 9.1 mph = 13.3 ft/sec due to the action of the retarding impact force.

If the car is shortened about 8" in this impact, then the distance through which the retarding force acts is about 12" (the truck starts moving during the impact; assume it moves about 4") so that the car

travels about 1' during the impact with an average velocity of about 8 ft/sec. Thus, the duration of the impact can be estimated as follows:
Distance = Velocity x Time
Time = Distance/Velocity
T = 1 ft/(8 ft/sec) = 0.125 sec
So the car decelerates from 14.7 ft/sec to 1.3 ft/sec in a time of 0.125 sec. Thus, its average deceleration is:
13.4 ft/sec / 0.125 sec = 107 ft/sec/sec = 3.33 g

The average force acting on the car then is 3.33 x 3,000# = 9990# This is also the force that acts on the truck (Newton's Third Law: Every force has an equal and opposite reaction force): So that the average acceleration of the truck is: 30,000/9990 = 3 ft/sec/sec = 0.1 g. For Accident Statistics Motor Vehicle Fatal Crashes by Day of Week, Time of Day, and Weather and Light Conditions (percent) [Source: Bureau of National Transportation Statistics] go to: http://www.car-accidents.org/stats-conditions.htm

Greg MacMaster — Environmental Forensics

QUESTIONS FOR DISCUSSION

1. A lawyer calls and asks for a temperature at noon in Dallas, TX on 3/19/2008. Discuss with class on questions you would ask concerning the lawyers' inquiry.

2. If a lawyer asks if he/she can use your name and add it to the list of expert witnesses they may call on, do you let them do it? What conditions would you consider before doing this? Would you charge money? What implications could arise by doing this? What advantage is this to the attorney?

3. How important is it to a case to visually see the geographic region or area before accepting the case? Explain your answer.

4. If you were asked to provide weather data and drafted an affidavit explaining your theory and submitted it to the lawyer and later you realized a tree line, which wasn't noted in an earlier deposition and subsequently the case was dismissed, what would you do? Contact the lawyer and correct your findings?

5. How important do you think it is to a case by visiting the location before submitting your findings to the court?

6. By hiring a local forensic meteorologist, they have the upper hand when it comes to picking away at your testimony. Often times you may be hired to evaluate another forensic meteorologist' work and find holes in their theory. Would you call your friend whom happens to be on the opposing council's list of witnesses and discuss the case? Discuss with class or with a visiting attorney.

Greg MacMaster   Environmental Forensics

Chapter 4

Weather and Estimated Time of Death
Pathologist/Medical Examiner & Meteorologist

One of the responsibilities of the medical examiner with help from pathology testing is to estimate the time of death and an accurate assessment is of great importance to police in narrowing down the list of suspects. It can allow police to pinpoint the time during which they need to find out what the suspects were doing and allows them to eliminate people who have an alibi for that period from there enquiries. Can you imagine the possible suspects on a timing window of 8 to 12 hours? What if we, as Forensic Meteorologists could help reduce that window to, say 4 hours or less?

On a hot day in August in the Great Lakes, the body of an elderly man was found in a ditch next to a country road. He was laying supine (flat on his back), dressed in overalls, a short-sleeved work shirt and socks without shoes. His abdomen and legs were covered by a blanket. There was extensive maggot infestation of the head and neck, partially obscuring a scalp defect of the back of his head. In addition, there was a ligature (necklace, rope or something to show trauma) encircling the neck. The skin of the upper chest, neck and head was markedly darkened. The rest of the body was not decomposed. There was no apparent blood on the ground surrounding the decedent (dead person).

The scalp defect was a 4 < cm gaping laceration with no injury to the underlying bone or brain. The ligature was a small towel tightly compressing the neck and knotted in the back. There was no rigor mortis and livor mortis was present was present on the posterior aspect of the body. The stomach contained fragments of sausage, brown liquid and yellow-white semi-solid food particles. Approximately a dozen maggots were collected and preserved in alcohol. A diagnosis of ligature strangulation was rendered. The manner of death was ruled as homicide.

The next day the Sheriff called the pathologist and asked about the time of death. The deputy was holding a man in custody that had been seen with the decedent two days prior to the body being found. The suspect was out of town on business the day before the victim's body was discovered. The suspect had good motive for the murder because of a soured business deal with the victim. Prior to making a formal arrest, the officer needed to make sure the postmortem interval (PMI) was consistent within two days. What should the pathologist tell the deputy about the PMI in this case?

One of the most frequently asked questions during a death investigation concerns the time of death. Unfortunately, determining the exact time of death from an examination of the body isn't possible. Numerous findings must be interpreted to give a reasonable estimate of the PMI. In this particular case, the time of death was a key element to the arrest because the suspect had an alibi for the day prior to the discovery of the body.

Differential decomposition occurred because a head injury caused an open wound. Blood is an excellent stimulus for flies to lay eggs and for maggot activity which accelerated decompositional changes of the head. Since the rest of the body hadn't decomposed, an estimation of the PMI was made by evaluating the area of least decomposition. Environmental temperature is the most important factor in determining the rate of decompositional change after death. The decompositional changes in the head and the lack of rigor mortis could occur in 10-20 hours in temperatures circa (approx.) 90 degrees F. Therefore, in this case, the appearance of the body suggested a PMI of less than 24 hours.

The maggots appeared freshly hatched and were also consistent with being less than 30 hours of age. This information correlates with other findings since it is well established that flies lay eggs on a body very shortly after death. This would suggest the decedent died the day before his body was discovered. Therefore the "prime" suspect with the strong alibi does not appear to have committed the murder, and the investigators need to renew their search for the killer.

To a Forensic Meteorologist (FM) who has an interest in this type of forensic science may look back and wonder how long the temperature was 90 F. The peak warming time of the day is usually Noon to 4 p.m., and the best time to get close to 90F. Would the temperature change the PMI? Would there be a measurable temperature change from the ground to 2 meters above the ground? What about absorption of the suns rays? Black Asphalt vs. Concrete will show diurnal variations, and during mid day when heating is the greatest, the concrete will be much cooler. All these factors, taken into consideration would change the PMI.

A difference in location yet the same time can have an impact of the decomposition of a body. Following death, the body begins to cool at a rate which depends on a range of factors. A naked body will cool faster than a clothed one; a large adult will cool slower than that of an infant; a body in a prone position will cool faster than a body slumped in the corner; a body exposed to air currents will cool faster than one in a protected area. Other factors such as topography, contact to water and other such factors will have an influence.

The body cools according to a defined set of formulas, which have been noted in this chapter. One parameter which has not been defined in flux and that is air temperature. In the following spreadsheet, I calculated the difference in the hours in relationship to the body and air cooling at different rates.

Air Temperature measured at time of first measurement - None taken in succession (does not factor in Heat Index or Windchill).
Result is in hours.

	90 F	84 F	80 F	74 F	70 F	64 F	60 F	54 F
Body=95	12.9	7.7	6.2	5	5.8	5.1	4.8	4.4
Body=90		19.9	13.9	10.3	10.7	9.2	8.4	7.6
Body=85			31.1	19.1	17.7	14.5	13	11.4
Body=80				32.1	26.1	20	17.5	15

Computations taken from http://www.pathguy.com/TimeDead.htm

Constant Factors in computation:
Body Weight=200lbs

Found: Moving air/Dry Body

Covering: 1-2 Thin Layers
Liver/Rigor/hypothermia: not observed

As a meteorologist, we know that terrain has a drastic influence in temperature recordings and the difference in sensor height to the body location (ground) can complicate the input variables in formulating the time of death. If our values are off by 10 degrees, the estimated time of death could vary from 17.7 hours to 31.1 hours! Although there are other visible factors to help narrow down the window of death, increasing our accuracy in temperatures would be an excellent start.

Not to complicate things but what do humans feel when we are outside? Since we're not in a shaded box (like a temperature sensor) we tend to feel warmer or cooler than what the actual temperature depicts. This is called "Feels Like" temperature. More commonly called Heat Index and Wind-chill. If the sun is beating down on us, will our temperature rise over time? Why wouldn't a dead body react the same way? Just because their sense of feeling has been denied doesn't mean it can't warm or cool accordingly.

In the table on the previous page, the air temperature did not include any factors which could hinder the standard cooling rate. With our abilities to adjust the temperature based on the Heat Index and Wind-chill formulas, can you see how varied the outcome would be? Now you can see why it is so imperative to give detailed weather information to the investigators when they have to determine time of death. This also applies to the scientific field of entomology (more emphasis on Heat Index whereas Wind-chill is not as much of a factor).

What follows is a formula for the Heat Index:
If you know the relative humidity and the dry air temperature, then you can use the following equation to calculate the heat index.

Heat index (HI), or apparent temperature (AI) = $-42.379 + 2.04901523(Tf) + 10.14333127(RH) - 0.22475541(Tf)(RH) - ((6.83783 \times 10^{-3})(Tf^2) - ((5.481717 \times 10^{-2})(RH^2) + ((1.22874 \times 10^{-3})(Tf^2)(RH)) + ((8.5282 \times 10^{-4})(Tf)(RH^2)) - ((1.99 \times 10^{-6})(Tf^2)(RH^2))$

Note: In order for the Heat Index formula to work correctly, you must use the relative humidity in percent form. In other words, if the

relative humidity is 65%, use 65 for RH in the formula, not .65.
Tf=Temperature if F.
This equation takes into account fairly light winds. Exposure to direct sunlight can increase the Heat Index by up to 15°F.
The reverse would be true in cooler seasons.
Specifically, the new Windchill index:

- Calculates wind speed at an average height of five feet (typical height of an adult human face) based on readings from the national standard height of 33 feet (typical height of an anemometer)
- Is based on a human face model
- Incorporates modern heat transfer theory (heat loss from the body to its surroundings, during cold and breezy/windy days)
- Lowers the calm wind threshold to 3 mph
- Uses a consistent standard for skin tissue resistance
- Assumes no impact from the sun (i.e., clear night sky).

The new formula for winds in mph and Fahrenheit temperatures is:

Wind chill temperature = 35.74 + 0.6215T - 35.75V (**0.16) + 0.4275TV(**0.16)

In the formula, V is in the wind speed in statute miles per hour, and T is the temperature in degrees Fahrenheit.

Note: In the formula, ** means the following term is an exponent (i.e. 10**(0.5) means 10 to the 0.5 power, or the square root of V), - means to subtract, + means to add. A letter next to a number means to multiply that quantity represented by the letter by the number. The standard rules of algebra apply.

In the fall of 2001, the U.S. National Weather Service and the Canadian weather replaced the old formulas with new ones. The new formulas are based on greater scientific knowledge and on experiments that tested how fast the faces of volunteers cooled in a wind tunnel with various combinations of wind and temperature.

As a courtesy to the National Weather Service, they have provided both scales to help determine Heat Index and Wind-chill. These can be found at the end of this book. In 1978, a case occurred in a remote part of New South Wales which illustrates this rather dramatically. The local telephone operator suspected foul play when she could not connect a call to a remote farmhouse after a number of attempts. Eventually a young child answered. After some questioning the child explained that she "did not like Mommy any more because she is turning black". When the police arrived they found two bodies in quite different states of decay even though they had been killed at the same time.

The father had been shot on the kitchen and lay dead on the cool linoleum floor. The mother was found shot dead in bed and was in a far more advanced state of decomposition because the electric blanket had bee left on. Because the body had been kept warm the bacteria found naturally in the human body had multiplied rapidly and begun the decay process. Flies had laid eggs in the body and maggots had hatched and grown. The state of the woman's corpse had deteriorated rapidly, while very little change occurred in the male victim. Meteorologists can grasp this concept and have a really good idea how to deal with the use of thermodynamic formulas. First we'll touch on a single variable (such as outside temperature) in the following formula.

An Application of Newton's Law of Cooling
Police arrive at the scene of a murder at 12 am. They immediately take and record the body's temperature, which is 90F, and thoroughly inspect the area. By the time they finish the inspection, it is 1:30 am. They again take the temperature of the body, which has dropped to 87F, and have it sent to the morgue. The temperature at the crime scene has remained steady at 82F. When was the person murdered? Break the problem into subsections;

Let $T(t)$ be the temperature of the body at time t; take t=0 to be 12 midnight. We have this information.

Time	Body Temp.	Ambient Air Temp.
0 hr	90	82
1.5 hr	87	82

Greg MacMaster — Environmental Forensics

Here, Newton's law of cooling is: $dT/dt = k*(82-T)$,
where k is an unknown constant.

This is a separable ODE. Solve it in these steps.
1. Put it in differential form. $(T-82)^{-1} dT = -k*dt$
2. Integrate both sides. $\ln|T-82| = -k*t + C$ (Implicit form of solution.)

Next, we need to find k and C in step 2.

From the table, we get

$\ln|90 - 82| = -k*0 + C$ (T(0)=90)
$\ln|87 - 82| = -1.5*k + C$ (T(1.5)=87)

Simplifying and solving these, we get C=ln(8) and k=ln(8/5)/1.5.

Inserting these in the implicit form of the solution in step 2, and using T(tD)=98.6 at the time of death, we have

$\ln|98.6 - 82| = -tD*\ln(8/5)/1.5 + \ln(8)$
so
tD = - 1.5*ln(16.6/8)/ln(8/5) = - 2.3296, about 9:40 pm. To have an idea how temperature works on the body, we have to know what the formulas are and how the data is entered in and learn the basics of decomposition over time.

(Referenced from the Medical Algorithms Project. Developed by the Institute for Algorithmic Medicine , a non-profit Texas Corporation.

Overview of formulas: The temperature of a body can be used to estimate the time of death.

Method 1
time since death in hours = ((rectal temperature at time of death in °F) - (rectal temperature in °F at time t)) / 1.5
Method 2

time since death in hours = ((rectal temperature at time of death in °C) - (rectal temperature in °C at time t)) + 3
Limitations:

- Body temperatures at the time of death that are above or below normal can give misleading results.

References: Green MA Wright JC. Postmortem interval estimation from body temperature data only. Forensic Science International. 1985; 28: 35-46. Knight B Nokes L. Chapter 2: Temperature-based methods I. pages 3-45 (pages 34-35). IN: Henssge C Knight B et al. The Estimation of the Time Since Death in the Early Postmortem Period. Arnold Publishers. 1995.

Equation of De Saram
Overview: The temperature of a body can be used to estimate the time of death.

time since death in hours = ((((rectal temperature at time of death in °F) - (rectal temperature in °F at time t1)) / ((rectal temperature at time t1 in °F) - (rectal temperature in °F at time t2))) * (t2 - t1)) + 0.75
where:
- 0.75 hours (45 minutes) is a factor that can be added to take into account any delay in cooling

Limitations:
- Body temperatures at the time of death that are above or below normal can give misleading results.

References:
De Saram G Webster G Kathirgamatamby N. Post-mortem temperature and the time of death. J Crim Law Criminol Police Sci. 195; 1: 11-24.
Knight B Nokes L. Chapter 2: Temperature-based methods I. pages 3-45 (page 35). IN: Henssge C Knight B et al. The Estimation of the Time Since Death in the Early Postmortem Period. Arnold Publishers. 1995.

Method of Green and Wright

Overview: The method of Green and Wright for estimating the time of death uses rectal and environmental temperatures. It assumes that a double exponential mode adequately describes how the body cools.

gradient G = ((rectal temperature in °C at time 2) - (rectal temperature in °C at time 1)) / (time interval in hours between time 1 and time 2

where:

- The original equation shows temperature at time 1 minus temperature at time 2 but states that G is always negative since the latter temperature is always greater than the former.

gradient A = (((rectal temperature in °C at time 1) + (rectal temperature in °C at time 2)) / 2) - (average environmental temperature in °C)

reduced theta = ((rectal temperature at the time of death) - (((rectal temperature in °C at time 1) + (rectal temperature in °C at time 2)) / 2)) / ((rectal temperature in °C at time of death) - (average environmental temperature in °C))

where:
- The rectal temperature at the time of death is assumed to be 37°C.

Estimation of F

For reduced theta between 0 and 0.639 the rise in F is linear and is approximated by the equation derived in JMP:

F = (1.8826 * (reduced theta)) - 0.026636

For reduced theta from 0.639 to 0.96 the rise in F is exponential and can be approximated by the equation derived in JMP.

F = (317.62648 * ((reduced theta)^4)) - (926.6809 * ((reduced theta)^3)) + (1014.8223 * ((reduced theta)^2)) - (491.3695 * (reduced theta)) + 89.622192

Calculation of Time Since Death

time since death in hours = (-F) * ((gradient A) / (gradient G))

Limitations:

- The method assumes that the rectal temperature at the time of death was normal which may or may not be an accurate assumption.
- The method assumes that the environmental temperature remains relatively constant over the time since death.

References:
Green MA Wright JC. Postmortem interval estimation from body temperature data only. Forensic Science International. 1985; 28: 35-46.
Green MA Wright JC. The theoretical aspects of the time dependent Z equation as a means of postmortem interval estimation using body temperature data only. Forensic Science International. 1985; 28: 53-62.
Knight B Nokes L. Chapter 2: Temperature-based methods I. pages 3-45 (pages 22 and 38). IN: Henssge C Knight B et al. The Estimation of the Time Since Death in the Early Postmortem Period. Arnold Publishers. 1995.

Method of Fiddes and Patten

Overview: The method of Fiddes and Patten for estimating the time of death uses rectal and environmental temperatures. It uses the virtual cooling time which is the time it takes a body to reach a 15% difference between rectal and environmental temperatures.

Variables:
(1) rectal temperature in °C at the time of death (normally 37
(2) rectal temperature in °C at time 1
(3) rectal temperature in °C at time 2
(4) time interval in hours between time 1 and time 2
(5) environmental temperature in °C at time of death

percent fall in rectal temperature at time 1 = ((rectal temperature in °C at time of death) - (rectal temperature at time 1)) / ((rectal temperature in °C at time of death) - (environmental temperature in °C))

percent fall in rectal temperature at time 2 = ((rectal temperature in °C at time of death) - (rectal temperature at time 2)) / ((rectal temperature in °C at time of death) - (environmental temperature in °C))

Estimating Percent of Virtual Cooling Time from Percent Fall in Temperature Difference

The percent fall in rectal temperature correlates with the percent fall in the virtual cooling time in a semilogarithmic manner (fall in temperature semilogarithmic percent virtual cooling time linear). Using data in Hennsge (1995 Figure 2.22 page 36) the fall in temperature difference and virtual cooling time can be approximated by equations from JMP:

For percent fall in temperature from 0 to 30 percent
percent of virtual cooling time = (0.00375 * ((percent fall in temperature) ^ 2)) + (0.5825 * (percent fall in temperature)) + 0.075

For percent fall in temperature from 30 to 60 percent
percent of virtual cooling time = (0.005 * ((percent fall in temperature) ^ 2)) + (0.49 * (percent fall in temperature)) + 1.7

For percent fall in temperature from 60 to 85 percent
percent of virtual cooling time = (0.0546231 * ((percent fall in temperature) ^ 2)) - (5.972111 * (percent fall in temperature)) + 211.30905

These are calculated for the percent fall in temperature seen at time 1 and time 2.

Estimating the Time Since Death

difference in percent virtual cooling time between time 1 and time 2 =
= (percent of virtual cooling time at time 2) - (percent of virtual cooling time at time 1)

100% of virtual cooling time in hours = (time interval in hours between time 1 and time 2) * 100 / (difference in percent virtual cooling time between time 1 and time 2)
time since death in hours = (100% of virtual cooling time in hours) * (percent fall in rectal temperature at time 1)

Limitations:
• The precise rectal temperature at the time of death may not be 37°C. The more that the actual rectal temperature deviates from this value then the more inaccurate the estimates.
• The environmental temperature needs to be relatively constant for the method to provide meaningful estimates.

References:
Knight B Nokes L. Chapter 2: Temperature-based methods I. pages 3-45 (pages 20-21 and 35-36). IN: Henssge C Knight B et al. The Estimation of the Time Since Death in the Early Postmortem Period. Arnold Publishers. 1995. (NOTE: It appears as if the legends for Figure 2.22 page 36 are switched. This would correlate with Figure 8 page 21 and would make more sense with the text.)

Method of Al-Alousi and Anderson

Overview: The method of Al-Alousi and Anderson to estimate the time of death uses organ and environmental temperature to calculate a temperature difference ratio which is then compared to a cooling curve. The method uses different cooling curves depending on whether the body is clothed or naked.
Organ temperature to measure:
(1) rectum
(2) liver
(3) brain
(4) any organ for which a cooling curve is available
(5) Thigh or other "meaty" part of the body where a steady temperature reading can be obtained.

temperature difference ratio = ((organ temperature in °C at time 1) - (environmental temperature in °C at time 1)) / ((organ temperature in °C at time of death) - (environmental temperature in °C at time 1))

Liver Body Naked

For temperature difference ratio >=0.23
mean time since death in hours at time 1 = (-39.477 * ((temperature difference ratio)^3)) + (104.578 * ((temperature difference ratio)^2)) - (103.96 * (temperature difference ratio)) + 38.859

For temperature difference ratio >=0.06
mean time since death in hours at time 1 = (863.727 * ((temperature difference ratio)^2)) - (405.879 * (temperature difference ratio)) + 67.726

For ratio < 0.06 time of death is assumed to be > 50 hours.

Liver Body Covered

For temperature difference ratio >=0.5
mean time since death in hours at time 1 = (-40 * (temperature difference ratio)) + 40

For temperature difference ratio >=0.17
mean time since death in hours at time 1 = (163.487 * ((temperature difference ratio)^2)) - (191.52* (temperature difference ratio)) + 75.143

For ratio < 0.17 time of death is assumed to be > 50 hours.

Rectum Body Naked

For temperature difference ratio >=0.22
mean time since death in hours at time 1 = (29.677 * ((temperature difference ratio)^2)) - (61.847 * (temperature difference ratio))+32.17

For temperature difference ratio >=0.01
mean time since death in hours at time 1 = (320.2587 * ((temperature difference ratio)^2)) - (205.8496 * (temperature difference ratio)) + 49.8496

For ratio < 0.01 time of death is assumed to be > 50 hours.

Rectum Body Covered

For temperature difference ratio >=0.33
mean time since death in hours at time 1 = (27.4789 * ((temperature difference ratio)^2))-(66.3977 * (temperature difference ratio)) + 38.91878

For temperature difference ratio >=0.01
mean time since death in hours at time 1 = (211.783 * ((temperature difference ratio)^2)) - (159.305 * (temperature difference ratio)) + 49.59259

For ratio < 0.01 time of death is assumed to be > 50 hours.

Limitations:

- The body temperature at the time of death may not be 37 °C.
- If the environmental temperature varies widely and is not constant then the environmental temperature used may be misleading.
- Above 40-50 hours the curves flatten and may not be accurate.
- The cooling curves show a hatched area +/- 1 standard deviation so that the estimates are only rough time indicators.

References:
Knight B Nokes L. Chapter 2: Temperature-based methods I. pages 3-45 (pages 23 - 24 and 38). IN: Henssge C Knight B et al. The Estimation of the Time Since Death in the Early Postmortem Period. Arnold Publishers. 1995

Most, or all, of the computer programs on the market are asking for a single OAT [outside air temperature] (no change over time). As a forensic meteorologist, I have rarely seen or recorded the same outside air temperature for more than a few hours. Which leads me to believe that our timing error in computing time of death, is due to the nonexistence of a non-linear differential formula, to account for the adjusted OAT over time.

A new computer program is currently being tested which takes all of these formulas into account along with non-linear algorithms to create a more realistic regression of temperature over time and introduces other facets of forensic evidence to help narrow down the Estimated Time of Death (ETOD). Unlike the versions that are currently running where it's derived by body temperature alone. This new program takes into account other factors in the PMI. For more information on this program and updates to ETOD, go to:

www.forensicweatherman.com or
http://groups.yahoo.com/group/forensic_meteorology/
Facebook Group: Environmental Forensics and Its affects on Investigations

The timing of death is both an art and a science. Unless the death is witnessed, it is impossible to determine the exact time of death. It is assumed that if someone knows the time of death, that they should be considered a suspect. The Medical Examiner (ME) can only "estimate" the approximate time of death and it is important to note that this "estimated time of death" can vary greatly from the "legal" time of death, which is the time recorded on the death certificate, or the "physiologic" time of death, which is when vital functions actually cease. The "legal" time of death is the time the body was discovered or the time a doctor or other qualified person pronounced the victim dead. These "times of death" may differ by days, weeks, even months, if the body is not found until well after "physiologic" death has occurred.

For example, if someone kills a victim in July, but the body is not discovered until October, the "physiologic" death took place in July, but the "legal" death is marked as October. The Medical Examiner's "estimated time of death" would be July. The ME can estimate the "physiologic" time of death with some degree of accuracy. He uses the decomposition changes that occur in the human body after death to help him in this endeavor. These changes consist of measuring the drop in body temperature, the degree of rigidity (rigor mortis), the degree of discoloration (livor mortis or lividity), the stage of body decomposition, stomach contents, and other factors. Bodies found in water present special problems in this regard.

A feasibility study is underway to determine if EMS personnel can assist in determining the time of death in areas where it's physically impossible for the Medical Examiner to attend the crime scene. Protocol standards will have to be written and approved by the Prosecuting Attorney and fall in line with the National Guidelines of Death investigations. This would be a benefit for those areas that are not fully staffed with enough Medical Examiners to investigate every crime scene due to distance or other limiting factors. The result: More accurate assessment of when the crime occurred.

Body Temperature:
Normal body temperature during life is 98.6 degrees F. This will vary if a body was attacked or there was physical activity prior to death, then the body temperature would be higher (in some cases as much as 2-4 degrees). After death, the body loses heat progressively until it equilibrates with that of the surrounding medium. The rate of this heat loss is approximately 1 to 1.5 degrees per hour until the environmental temperature is attained, then it remains stable. Obviously, this measure is greatly affected by location and clothing. A body in the snow in Michigan in January and one in a Louisiana swamp in August will lose heat at widely divergent rates. These factors must be considered in any "estimate" of time of death. Even

if the body temperature is equal to the OAT, you can reference hourly temperature records to determine how long the OAT was equal to the body temperature. Remember, the OAT can warm or cool faster than the body. (unless water is a factor).

The criminalist who processes the scene should take a body temperature and measure the temperature of the surrounding medium--air, water, snow, earth (if the body is buried). Ideally, the body temperature is taken rectally or through the liver. Obviously, the sooner after death the body is found, the more accurately time of death can be assessed by this method. Once the body reaches ambient temperature, all bets are off. Air temperature at 2 meters isn't sufficient because the body is on the ground, so take temperatures that are relative to the surroundings. Keep in mind that daytime temperatures where the sun shines will differ to shaded readings (such as the standard readings obtained by official standards).

Rigor Mortis:
Rigor mortis is the stiffening and contraction of the muscles due to chemical reactions that take place in the muscle cells after death. It typically follows a predictable pattern. Rigidity begins in the small muscles of the face (cheek area) and neck and progresses downward in a "head-to-toe" fashion to the larger muscles. The entire process takes about 8-12 hours. At that time, the body is completely stiff and is "fixed" in the position of death. Then, the process reverses itself, with rigidity being lost in the same fashion, beginning with the small muscles and progressing to the larger ones. This process begins 18 to 36 hours after death and is usually complete within 48 hours. So, rigor is only useful in the first 48 to 60 hours after death. Physical activity will speed up the process.

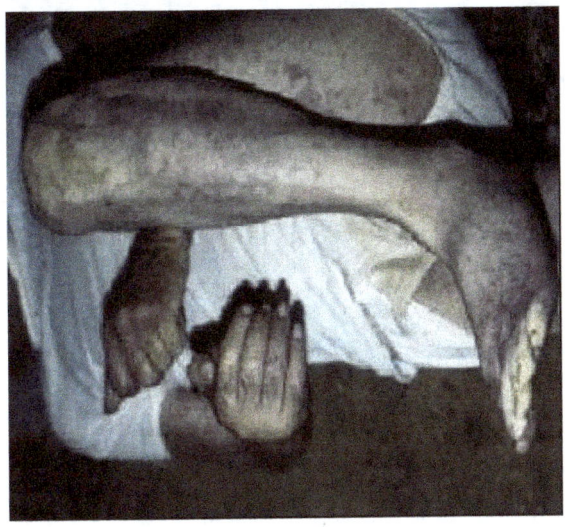

(Rigor Mortis (Photo Courtesy of Jay Dix and Michael Graham – Time of Death, Decomposition and Identification, an Atlas.)

The reason for the rigidity is the loss of adenosine triphosphate, or ATP, from the muscles. ATP is the compound that serves as energy for muscular activity and its presence and stability depend upon a steady supply of oxygen and nutrients, which are lost with the cessation of cardiac activity. The later loss of rigidity and the appearance of flaccidity (relaxation) of the muscles occur when the muscle tissue itself begins to decompose. Rigor is one of the least reliable methods for determining time of death because it is extremely variable. Heat quickens the process, while cold slows it. Obese people may not develop rigor, while in thin victims it tends to occur rapidly. If the victim struggled before death and consumed much of his muscular ATP, the process is hastened. Exercise and heat also play a roll in speed of rigor. After a 24 mile marathon, the body could be fully rigor within 6 hours because it can't replace the oxygen that was lost during exercise.

Simplified table of Rigor Mortis

Temp of body	Stiffness of body	Time since death
Warm	Not stiff	Not dead more than three hours
Warm	Stiff	Dead between 3 to 8 hours
Cold	Stiff	Dead between 8 to 36 hours
Cold	Not stiff	Dead in more than 36 hours

Lividity:
Lividity is caused by stagnation of blood in the vessels. It lends a purplish color to the tissues. The blood, following the dictates of gravity, seeps into the dependent parts of the body--along the back and buttocks of a victim who is supine after death. Initially, this discoloration can be "shifted" by rolling the body to a different position, but by 6 to 8 hours, it becomes "fixed." If a body is found face down, but with fixed lividity along the back, then the body was moved at least 6 hours after death, but not earlier or the lividity would have "shifted" to the newly dependent area.

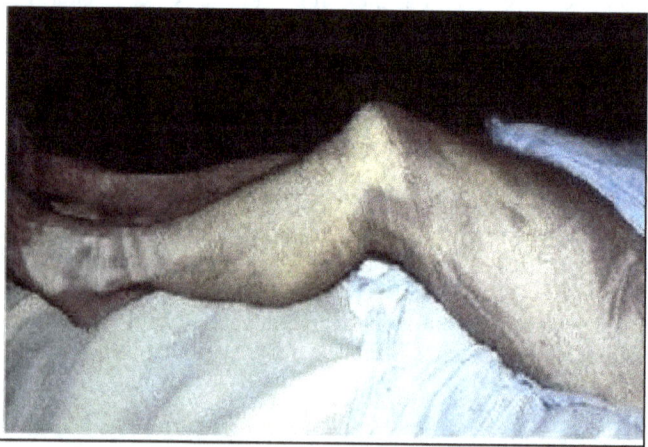

Lividity (Photo Courtesy of Jay Dix and Michael Graham – Time of Death, Decomposition and Identification, an Atlas.)

Body Decomposition:
At death, the body begins to decompose. Bacteria go to work on the tissues and by 24 to 36 hours the smell of rotting flesh appears and the skin takes on a progressive greenish-red color. By 3 days, gas forms in the body cavities and beneath the skin, and may leak fluid and split. From there, it gets worse. Add to this, predation by animals and insects and the body can become completely skeletonized before long. In hot, humid climes, this can happen in 3 or 4 weeks.
Stomach Contents:

The ME can often use the contents of the victims stomach to help determine time of death. After a meal, the stomach empties itself in approximately 4 to 6 hours, depending on the type and amount of food ingested. If a victim stomach contains largely undigested food material, then the death likely occurred within an hour or two of the meal. If the stomach is empty, the death likely occurred more than six hours after eating. Additionally, if the small intestine is also empty, death probably occurred some 12 hours or more after the last meal.

This table describes the process in the decomposition of the pig.

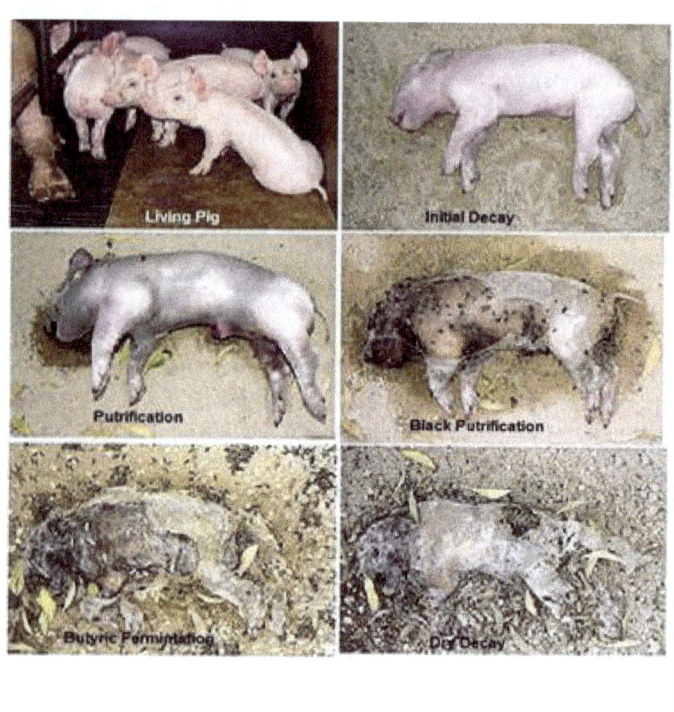

Stage	Description
Initial Decay	The cadaver appears fresh externally but is decomposing internally due to the activities of bacteria, protozoa and nematodes present in the animal before death
Putrefaction	The cadaver is swollen by gas produces internally, accompanied by odor of decaying flesh
Black Putrefaction	Flesh of creamy consistency with exposed parts black. Body collapses as gases escapes. Odor of decay very strong

Butyric Fermentation	Cadaver drying out. Some flesh remains at first, and cheesy odor develops. Ventral surface moldy from fermentation
Dry Decay	Cadaver almost dry; slow rate of decay

Floaters:
"Floaters" are corpses found floating in a body of water. They present special problems for the ME in determining the time of death. Water temperature of course has an effect as do local tides and predators. The general rule regarding decomposition is that one week on dry land equals two weeks for a submerged body.

To become a "floater," a body must to be in the water long enough for tissue decomposition from bacteria to begin. This process forms gas as a byproduct, which collects beneath the skin and in body cavities. Bodies tend to sink, then rise again in several days when the gas forms, adding buoyancy. They thus become "floaters."

Under these circumstances, the hands and feet swell (several days), the outer layer of skin separates from the underlying tissues (5-6 days), the skin of the hands and the nails separate (8-10 days), and entire body swells shortly thereafter. Tissues become extremely fragile and are easily damaged during removal from the water. Timing of the "floating" depends upon several factors, including water temperature, currents, the size of the victim, and other variables. For example, a body will "float" after 8-10 days in warm water and 2-3 weeks if in colder water. Cold slows the process of decomposition.

As you can see, the timing of a victim's death is a very inexact science and is greatly altered by the environment. In cold areas, body temperature changes are magnified, but decomposition changes are slowed, the inverse is true for hot, humid climes. Add to this, predation by insects and animals and the ME's job can become difficult. (DPLyle,MD 2000).

However, with the use of a forensic meteorologist (FM), the timing error can be greatly reduced offering the investigators a smaller timing error or more information to help close the investigation

The same theory holds true for entomologist (Study of insects). Tests have been done to determine rate of growth from egg to adult fly. However, these tests were performed at a constant temperature. Since insects are often found outside, how often do we observe a constant temperature for 12 hours? How about 24, 36 or 48 hours? It can't happen! Temperatures change and so does the growth rate of insects. Thus, a timing error is often incorporated due to the inexact science of fly-rearing. Is there a way to reduce the timing error? I believe so and it would take the use of sample data to compare against a Non-linear differential equation to find the answer.

Now let's apply what we have learned using temperature to determine

Time of Death:

You are the death investigator at a scene where a 47 year old man has been found dead on the floor of his bedroom. You arrive at midnight, and immediately obtain a core rectal temperature of 92.4° F. Room temperature is 70° F. Rigor is present in the jaw, arms, hands, and feet. Livor is well-established but is not fixed. The man is dressed in his pajamas, and his height and weight as stated on his drivers' license are 6'3" and 195 pounds. These are consistent with his appearance.

1. Using the rough calculation where the body cools at 1.5 degrees per hour, what do you estimate the time of death to be?
2. You learn that the thermostat in the house was programmed to change at 11 PM from 76° F to 70° F. Would this change your estimate of time of death using method #1?
3. You learn that the thermostat in the house was programmed to change at 11 PM from 76° F to 70° F. Would this change your estimate of time of death using method #2?

If you noticed, the formulas asked for a single temperature. What if the temperature changed over time? Would that have an influence in the calculation? Back in the beginning of this book I had some numbers from my thermostat;

Main floor thermostat is located on main floor (5 feet high)

Second story bed	70.7
Main floor wall	69.7
Placed on carpet /main floor	67.9
Basement floor	65.7

There's a 5 degree difference between the basement floor and the second story bed. Would this 5 degree difference change the ETOD? Can you imagine the offset if we used temperatures outside? It's imperative that we give the best possible dataset to aid investigators.

Eyes
Within minutes the cornea films over, and the white of the eye goes grey. After around two hours the cornea goes cloudy, and within a day or two it goes opaque. On the third day the gas makes the eyes bulge. With advanced decomposition, the eyes retract.

Food in the stomach
(Found at autopsy)
A light meal is out of the stomach within 1 ½ - 2 hours
A medium meal is out of the stomach within 3 - 4 hours
A heavy meal is out if the stomach within 4 - 6 hours
There are variations: Liquid is digested faster than semi-solid food, which is digested fasted than solid food. Emotional state may also influence the rate of stomach emptying.

Vitreous Potassium
There is potassium in the body's intercellular fluid - much more than there is in the plasma, on the other side of the cell membrane. After death it starts to leak out so there is the same amount on either side of the membrane. This happens at a nice steady rate, allowing time of death to be established. Samples are normally taken from the vitreous fluid of the eye.

Insects
- Body Lice: they usually outlive their host by 3 - 6 days.
- Various insects: They like to lay their eggs on very fresh corpses. The eggs hatch out within 8-14 hours. After another 8-14 hours it sheds its skin and emerges as a bigger larva. This process is repeated several times, taking 10-12 days in total. People who know about insects can therefore look at the larva, see which stage it is at, and work out time of death.

Samples should be taken and preserved and given to an entomologist.
Plants
- Grass/plants beneath an object wilt, turn yellow or brown and dies. The rate depends on type of plant, season, climate, etc.
- Seasonal plants or remnants may help indicate a range of time.

Samples should be collected and shown to a botanist.
Putrefaction

Purification begins after about 2 days. The process is faster in damp places or when the body is exposed to air. Decay is about eight times faster in the air than underground. Too cold or too hot and the process won't happen. In very hot temperatures the body will dry out and mummify instead.

People with a lot of fat will decay faster. People who died of bacterial disease will also decay faster. However, some poisons preserve the body.
2-3 days: green staining begins on the right side of the abdomen.

Body begins to swell.
3-4 days: staining spreads. Veins go "marbled" - a brown-black discoloration
5-6 days: abdomen swells with gas. Skin blisters
2 weeks: abdomen very tight and swollen.
3 weeks: tissue softens. Organs and cavities start bursting. Nails fall off.
4 weeks: soft tissues begin to liquefy. Face becoming unrecognizable
4-6 months: formation of adipocerous, if in damp place. This is when the fat goes all hard and waxy.

Anamnestic Evidence:
This means evidence taken from the victim's daily habits. For example, three days' uncollected newspapers would suggest he has been dead three days. If he missed an appointment on a particular day then he was probably dead then. If all his food in the refrigerator is rotten and horrid it suggests he's been dead a while.

Greg MacMaster Environmental Forensics

Summary:
As morbid as it sounds, it's very interesting! As you can see, there are many variables that can assist in determining the estimated time of death. Depending on how long the body was undiscovered will determine the type of science used. Temperature is more prevalent when the body is warmer than the surrounding air temperature. After that, it becomes necessary to rely on other factors collectively. Temperature can be applied to the entomology portion of the investigation which is described in the next chapter. So the next time you are asked for weather data, learn more about the purpose and why the need for the request. This will give you a greater understanding and help them as well.

Sample case:
An 82 yr old female sitting in a chair fully clothed found dead of respiratory arrest – they think.

Last seen at 23:30 pm in same position
Pronounced dead at 07:30 next day
Moderate rigor
Body dry
Air temp when found 72F
Rectal Temp: 90.2F

Have students use different temperatures and compute how far back in time to when she died using 98.6F as a stop temperature.
Compare your findings with the rest of the class.
Would this time difference change possible suspects if it was a murder? Use whatever means you would have to determine what the low temperature would be (basic cooling curve). Assume no fronts or change in weather patterns.

Questions for Discussions

1. A doctor requests a high and low temperature over a period of a week for a location – what questions do you ask in return?

2. What implications, if any, could happen if weather information you passed on was too basic to the investigation?

3. How much of a difference could there be between an airport (certified weather sensor) temperature report 25 miles away from a wooded area in a valley by a stream? Why?

4. How would you go about conducting a study to determine temperature variations between two different locations, one of which is a certified ASOS site? Would you provide the corrected values of your finding s to the Medical Examiner? Why?

5. Discuss how moisture plays a roll in the cooling rate of a body.

6. In a classroom discussion, discuss how wind-chill affects the cooling rate of a body, if any.

7. Invite the local Medical Examiner to your class (or visit their place of business) and ask them what cases they remember that were temperature dependant.

Chapter 5

Forensic Entomology

Environmental Impacts in Estimating time of Death.

Forensic (or medico-legal) entomology is the study of the insects associated with a human corpse in an effort to determine elapsed time since death. Insect evidence may also show that the body has been moved to a second site after death, or that the body has been disturbed at some time, either by animals, or by the killer returning to the scene of the crime. However, the primary purpose of forensic entomology today is to determine elapsed time since death.

Forensic entomology was first reported to have been used in 13th Century China and was used sporadically in the 19th Century and the early part of the 20th Century, playing a part in some very major cases. However, in the last 15 years, forensic entomology has become more and more common in police investigations. Most cases that involve a forensic entomologist are 72 h or more old, as up until this time, other forensic methods are equally or more accurate than the insect evidence. However, after three days, insect evidence is often the most accurate and sometimes the only method of determining elapsed time since death.

After the initial decay, and the body begins to smell, different types of insects are attracted to the dead body. The insects that usually arrives first is the Diptera, in particular the blow flies or Calliphoridae and the flesh flies or Sarcophagidae. The females will lay their eggs on the body, especially around the natural orifices such as the nose, eyes, ears, anus, penis and, depending on the type of assault, the vagina. If the body has wounds the eggs are also laid in such. Flesh flies do not lay eggs, but deposits larvae instead.

After some short time, depending on species, the egg hatches into small larvae. This larvae lives on the dead tissue and grows fast. After a little time the larva molts, and reaches the second larval instar. Then it eats very much, and it molts to its third instar. When the larvae are fully grown it becomes restless and begins to wander. It is now in its prepupal stage. The prepupae then molts into a pupae, but keeps the third larval instars skin, which becomes the so-called puparium.

Typically it takes between one week and two weeks from the egg to the pupae stage. The exact time depends on the species and the temperature in the surroundings. A table of life histories to some species of blow flies and flesh flies:

Some development data on different species of blowflies (Calliphoridae) and fleshflies (Sarcophagidae) **(After Kamal, 1958)**

Life histories of 11 species of blowflies and fleshflies reared at 27 degrees Celsius, and 50 percent relative humidity.

Species	Total Immature (Days)	No. of Gen.	Egg (Hrs)	First instar (Hrs)	Second instar (Hrs)	Third instar (Hrs)	Prepupa (Hrs)	Pupa (Days)
Sarcophaga cooley	29	--	24	18	48	96	9	16
Sarcophaga sherma	28	--	22	16	48	104	8	14
Sarcophaga bullat	18	--	26	18	54	112	12	17
Phormia regina	23	16	18	11	36	84	6	11
Protophormia terr	27	15	17	11	34	80	6	11
Lucilia sericata	29	18	20	12	40	90	7	12
Eucalliphora lila	27	22	22	14	36	92	6	13
Cynomyopsis cadav	17	19	20	16	72	96	9	18
Calliphora vomit	5	26	24	48	60	360	14	23
Calliphora vicina	5	24	24	20	48	128	11	18
Calliphora terran	4	25	28	22	44	144	12	20

An illustration of the blowfly life cycle:

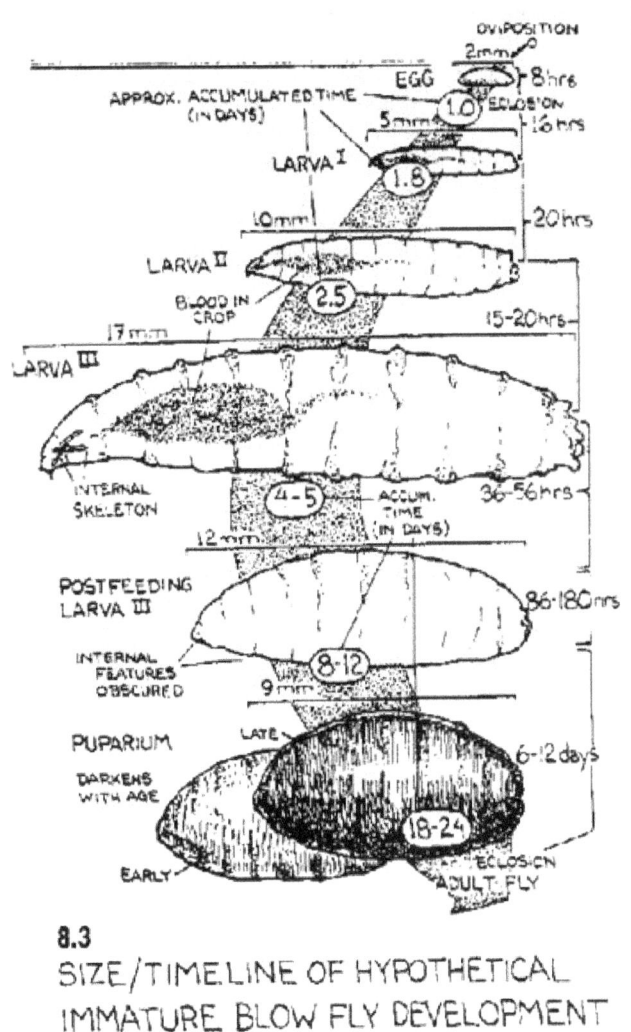

8.3
SIZE/TIMELINE OF HYPOTHETICAL IMMATURE BLOW FLY DEVELOPMENT

The theory behind estimating time of death, or rather the post mortem interval (PMI for short) with the help of insects are very simple: since insects arrive on the body soon after death, estimating the age of the insects will also lead to an estimation of the time of death. As you review the developmental stages, you'll see the temperatures that are fixed (not changing over time). We as meteorologist can start to see the benefit to our expertise and how it can change the developmental stage(s) of the insects.

How to estimate age of blowfly eggs, larvae, pupae and adults
Eggs:
When blow flies oviposit, their eggs has come very short in their development. The eggs are approximate 2 mm in length. During the first eight hours or so there are little signs of development. After that, and one can see the larvae through the chorion of the egg at the end of the egg stage. The egg stage typically lasts a day or so.

Larvae:
The blowfly has three instars of larvae. The first instar is approximately 5 mm long after 1.8 days, The second instar is approximately 10 mm long after 2.5 days, The third instar is approximately 17 mm long after 4-5 days. Identifying the right instar is the easiest part, and is done relatively easy based on size of larvae, the size of the larva's mouth parts and morphology of the posterior spiracles. The time it takes to reach the different instars depends very much on microclimate, i.e. temperature and humidity.

Prepupae:
At the end of the third instar the larva becomes restless and starts to move away from the body. The crop will gradually be emptied for blood, and the fat body will gradually obscure the internal features of the larvae. We say that the larva has become a prepupa. The prepupa is about 12 mm long, and is seen 8-12 days after oviposition.

Pupa:
The prepupa gradually becomes a pupa, which darkens with age. The pupa which are about 9 mm in length are seen 18-24 days after oviposition. The presence of empty puparia should therefore tell the forensic entomologist that the person in question has been dead in more than approximately 20 days. Identification can be done based on the remaining mouth parts of the third instar larvae.

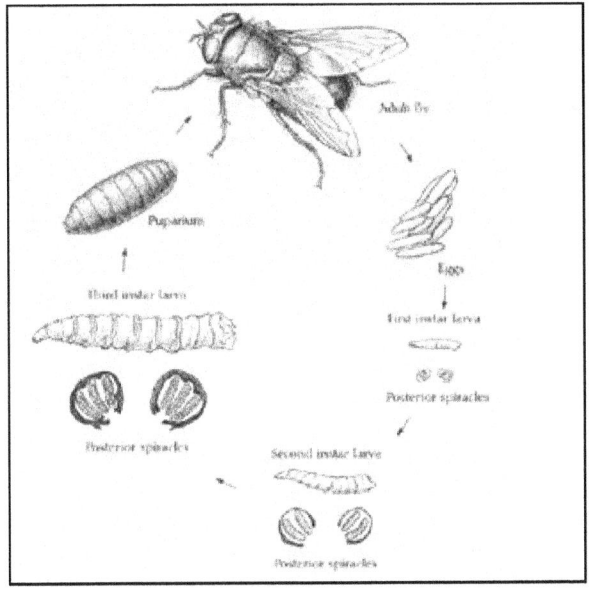

A more precise way to determine age of larvae and eggs is the use of rearing. For example: the body is found with masses of eggs on it, none have hatched. How long of a time is it since the eggs was oviposited? Note the time of the discovery, note the time when the first 1. Instar larvae occur. Subtract the first occurrence time with the discovery time, call this time A. Rear the blow flies to adults, let them mate, let them lay eggs on raw beef liver under conditions similar to the crime scene, take the time from oviposition to the first occurrence of 1. Instar larvae. Call this time B. By subtracting B-A, one gets C, which is an estimate of the time since oviposition to discovery. Similar calculations can be done for other instars as well. If one has good base-line data from before under different temperatures and for different species, one only needs to rear the flies to a stage where they can be identified, and that is the third stage or the adult stage.

One important biological phenomenon that occurs on cadavers is a succession of organisms that thrive on the different parts. i.e. beetles that specialize on bone will have to wait until bone is exposed. Predatory rove beetles or parasites that feed on maggots will have to wait until the blow flies arrive and lay their eggs. The succession on cadavers happens in a fairly predictable sequence and can be used in estimating time of death if the body has been lying around for some time.

Table of succession on guinea pigs from Bornemizza 1957

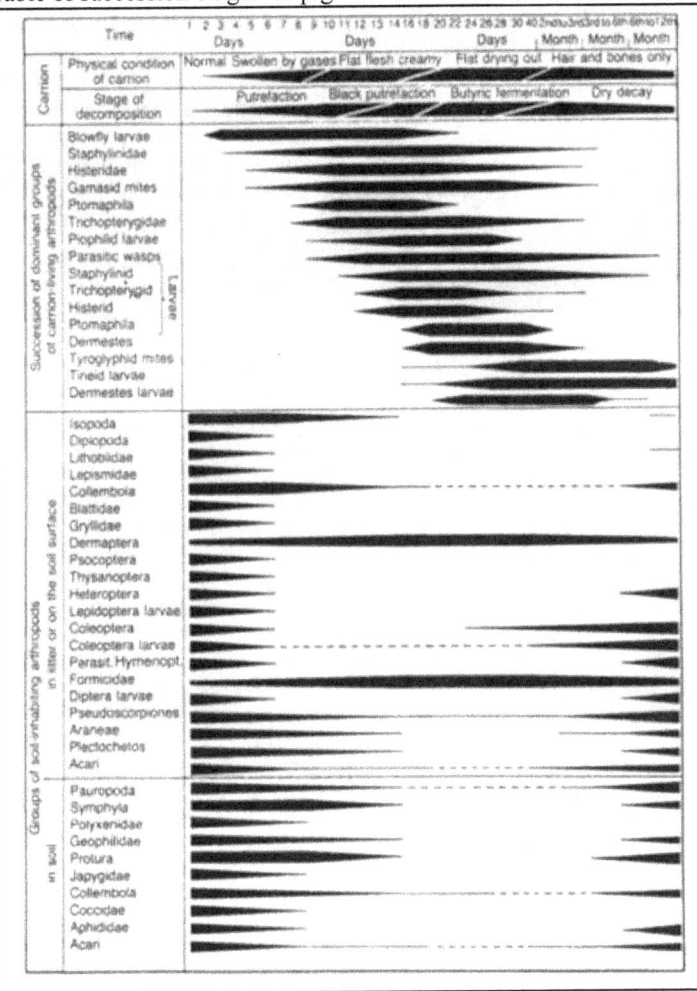

There are several things to note about this table:

The first group to arrive is blow flies, followed shortly by staphylinids. As putrefaction develops, more groups arrive at the scene, with most groups present just before the body is drying out due to seepage of liquids. After the body is drying out, dermestids, tineids and certain mites will be the dominant animal groups on the body, and blow flies will gradually vanish. Note also how the fauna changes in the soil. This can also be used to estimate time since death.

Succession data can be incorporated in a database, and when the forensic entomologist investigates a case, he can use the taxa found on the body as input, and get an estimate of the time of death as output. Several insects are specialized in living in very decayed dead bodies. One example is the cheese skipper, Piophila casei, where the larvae usually occur 3-6 months after death. The cheese skipper is a well known pest of cheese and bacon worldwide, and has a cosmopolitan distribution. Adult cheese skippers may occur early after death, but larvae occurs later.

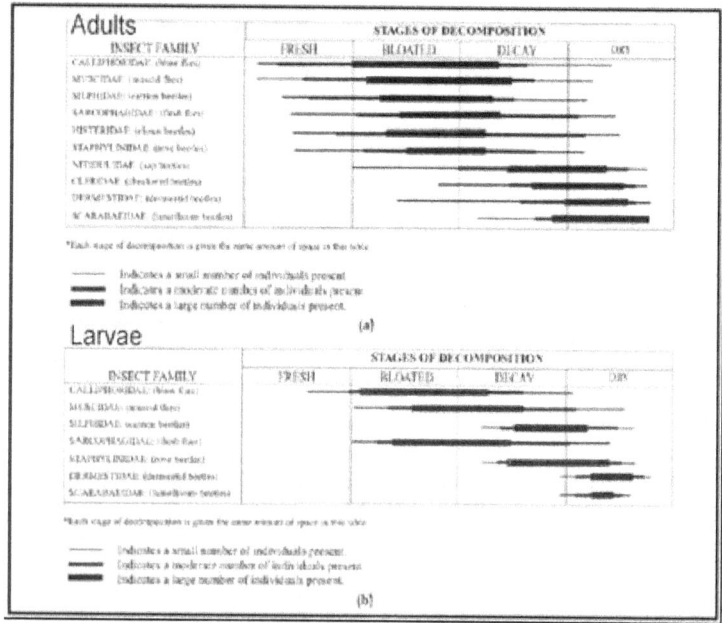

In temperate regions dead bodies often appear in spring, after the snow is gone. The forensic entomologist and the forensic pathologist must then try to determine whether the death occurred during the winter or before the snow set in. If the death occurred before

November, it is possible to find dead insects in and on the body. By analyzing the dead insect fauna and estimating when the insects probably died (this can be found by looking at meteorological records). Another hint is when the different adults stop flying before the winter. For example: There are some cases where the bodies have been found in the spring. In one case there were dead third stage blow fly larvae in the back of the mouth. The blow fly larva was of a species that is flying from May to October. It was from this conclusion that the eggs probably was laid during October, and since it was relatively few larvae, probably late in October. In another case, several live insects were found on a dead body, and also many dead third stage larvae.

The dead larvae were found on the stomach, the arms, the shoulders, and inside the head. A conclusion that the live insects had colonized the dead body in the spring and that the dead larvae had died during the winter. Based on the widespread occurrence of the larvae, it was likely that the body was colonized before October, probably in September.

If the death occurred in the winter things become difficult in outdoor settings, as very few insects are active in the winter. It is reported that larvae of the winter gnat, Trichocera sp. can develop on carrion in the winter. By estimating the age of these larvae, if present, it could be possible to estimate the PMI.

Although the amount of information discussed is well past the expertise of the meteorologist, it does give them a better idea as to the concept, application and effect the environment has on estimating time of death. If you are interested in knowing more about the collection and preservation of fauna, keep reading.

There are two main ways of using insects to determine elapsed time since death:

Using successional waves of insects
Using maggot age and development.
The method used is determined by the circumstances of each case. In general, the first method is used when the corpse has been dead for between a month up to a year or more, and the second method is used when death occurred less than a month prior to discovery.

The first method is based on the fact that a human body, or any kind of carrion, supports a very rapidly changing ecosystem going from the fresh state to dry bones in a matter of weeks or months depending on geographic region. During this decomposition, the remains go through rapid physical, biological and chemical changes, and different stages of the decomposition are attractive to different species of insects. Certain species of insects are often the first witnesses to a crime. They usually arrive within 24 h of death if the season is suitable i.e. spring, summer or fall in Canada and can arrive within minutes in the presence of blood or other body fluids.

These first groups of insects are the Calliphoridae or blowflies and the Muscidae or houseflies. Other species are not interested in the corpse when the body is fresh, but are only attracted to the corpse later such as the Piophilidae or cheese skippers which arrive later, during protein fermentation. Some insects are not attracted by the body directly, but arrive to feed on the other insects at the scene.

Many species are involved at each decomposition stage and each group of insects overlaps the ones adjacent to it somewhat. Therefore, with knowledge of the regional insect fauna and times of carrion colonization, the insect assemblage associated with the remains can be analyzed to determine a window of time in which death took place. This method is used when the decedent has been dead from a few weeks up to a year, or in some cases several years after death, with the estimated window of time broadening as time since death increases. It can also be used to indicate the season of death e.g. early summer. Knowledge of insect succession is required for this method to be successful.

The second method, that of using maggot age and development can give a date of death accurate to a day or less, or a range of days, and is used in the first few weeks after death. Maggots are larvae or immature stages of Diptera or two-winged flies. The insects used in this method are those that arrive first on the corpse, that is, the Calliphoridae or blowflies. These flies are attracted to a corpse very soon after death. They lay their eggs on the corpse, usually in a wound, if present, or if not, then in any of the natural orifices. Their development follows a set, predictable, cycle.

The insect egg is laid in batches on the corpse and hatches, after a set period of time, into a first instar (or stage) larva. The larva feeds on the corpse and moults into a second instar larva. The larva continues to feed and develop into a third instar larva.

The stage can be determined by size and the number of spiracles (breathing holes). When in the third instar, the larva continues to feed for a while then it stops feeding and wanders away from the corpse, either into the clothes or the soil, to find a safe place to pupate. This non-feeding wandering stage is called a prepupa. The larva then loosens itself from its outer skin, but remains inside. This outer shell hardens, or tans, into a hard protective outer shell, or pupal case, which shields the insect as it metamorphoses into an adult. Freshly formed pupae are pale in color, but darken to a deep brown in a few hours. After a number of days, an adult fly will emerge from the pupa and the cycle will begin again. When the adult has emerged, the empty pupal case is left behind as evidence that a fly developed and emerged.

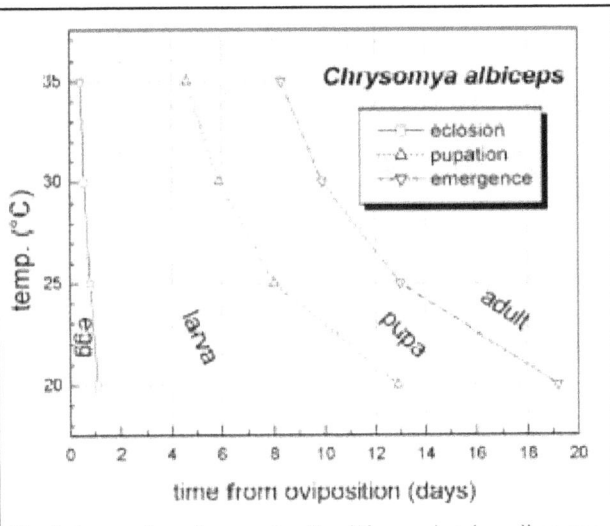

Fig. 3 Isomorphen diagram for *C. albiceps*, showing all stages from oviposition to eclosion. Areas between lines represent identical morphological stages at various temperatures. Each line represents identical morphological changes of this holometabolous insect; for values (mean days ± SD) see Table 1

Each of these developmental stages takes a set, known time. This time period is based on the availability of food and the temperature. In the case of a human corpse, food availability is not usually a limiting factor.

Insects are 'cold blooded', so their development is extremely temperature dependent. Their metabolic rate is increased with increased temperature, which results in a faster rate of development, so that the duration of development decreases in a linear manner with increased temperature, and vice-versa.

An analysis of the oldest stage of insect on the corpse and the temperature of the region in which the body was discovered leads to a day or range of days in which the first insects oviposited or laid eggs on the corpse. This, in turn, leads to a day, or range of days, during which death occurred. For example, if the oldest insects are 7 days old, then the decedent has been dead for at least 7 days. This method can be used until the first adults begin to emerge, after which it is not possible to determine which generation is present. Therefore, after a single blowfly generation has been completed, the time of death is determined using the first method, that of insect succession.

PROCEDURE

The first and most important stage of the procedure involved in forensic entomology involves careful and accurate collection of insect evidence at the scene. This involves knowledge of the insect's behavior; therefore it is best performed by an entomologist. Unfortunately, the entomologist is often not called until after the body has been removed from the death site. They usually see the remains at the morgue, and in some cases, do not actually see the remains at all, so their evidence is dependent on accurate collection by the investigating officers.

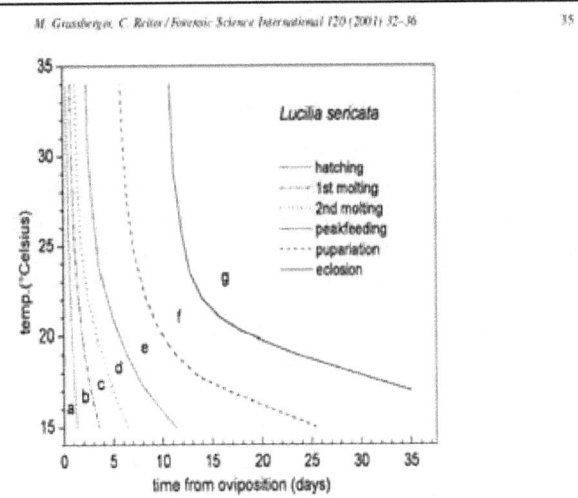

Fig. 3. Isomorphen-diagram for *L. sericata*, showing all stages from oviposition to eclosion (15–34 C). Areas between lines represent identical morphological stages at various temperatures. Where a = egg; b = 1st instar; c = 2nd instar; d = 3rd instar; e = postfeeding larva (i.e. prepupa); f = pupa; g = imago. Each line represents identical morphological changes of this holometabolous insect.

Table 9.1 Various Forensic Fly (Calliphoridae, Phoridae, and Sarcophagidae) Species with Their Corresponding Minimum Developmental Thresholds and Experimentally Determined Centigrade Degree-Day (CDD) and Centigrade Degree-Hour (CDH) Accumulations for Egg-to-Adult Emergence, Calculated from Data in Six Sources

Species	Dev. Min (°C)	Dev. Environ (°C)	Total Dev Time (h)	CDD	CDH	Data Ref. in Source
Source: Kamal (1958)[a]						
Calliphora terraenovae	6	26.7	551	475	11406	Table 1
Calliphora vicina	6	26.7	508	438	10516	Table 1
Calliphora vomitoria	6	26.7	854	737	17678	Table 1
Cynomyopsis cadaverina	6	26.7	439	379	9087	Table 1
Eucalliphora lilaea	10	26.7	330	230	5511	Table 1
Phaenicia sericata	10	26.7	348	242	5812	Table 1
Phormia regina	10	26.7	309	215	5160	Table 1
Protophormia terraenovae	10	26.7	301	209	5027	Table 1
Sarcophaga bullata	10	26.7	498	347	8317	Table 1
Sarcophaga cooleyi	10	26.7	402	280	6713	Table 1
Sarcophaga shermani	10	26.7	382	266	6379	Table 1
Source: Greenberg (1991)[b]						
Calliphora vicina	6	10	1647	275	6588	Table 3
	6	12.5	1069	290	6949	Table 3
	6	19	583	316	7579	Table 3
	6	23	460	364	8740	Table 3
	6	10	1626	279	6704	Table 7[a]
	6	12.5	1063	288	6910	Table 7[a]
	6	19	562	304	7306	Table 7[a]
	6	22	446	297	7136	Table 7[a]
			Mean	302	7239	

COLLECTING, PRESERVING AND PACKAGING SPECIMENS

Collection

Forensic Entomology is the use of the insects inhabiting decaying remains to determine certain forensic factors such as the postmortem interval (PMI) and is only successful when the relative evidence is properly collected and preserved. Collection equipment for this type of science isn't easy to find as most scientists usually put together a make-shift collection box of vials, jars, bags, forceps, marking utensils, small trowel and so on. An entomology kit should contain the following;

1 - <u>Entomology & Death</u> Book	1 - Small Hand Spade	1 - Pencil
1 - Flying Insect Swoop Net	1 - Hand Trowel	1 - Zippered Pencil Pouch
1 - 12" Tweezer	8 - Maggot Motel Screens	1 - Aluminum Foil - 25'
12 Pair of Latex Gloves	8 - 1 Pint Maggot Motels	1 - Camel Hair Brush
16 oz. - 70% Ethyl Alcohol	10 - 2 oz. Glass Jars	1 - Stainless Steel Spatula
1 - White 6" Photo Scale	4 - 8 oz. Plastic Jars	1 - Thermometer
1 - Photomacrographic Scale	10 - Glass Jars	1 - 1 oz. Plastic Bottle
1 - 3" Magnifier	12 - Small Boxes	1 - Hand Sanitizer
1 - Stainless Steel Pick	Entomology Labels	20 - Disposable Wooden Spatulas
50 - White 2" Adhesive Scales	Cotton Balls	1 - Checklist
50 - Biohazard Labels	1 - Roll Paper Towels	1 - Plastic Case with Foam Insert

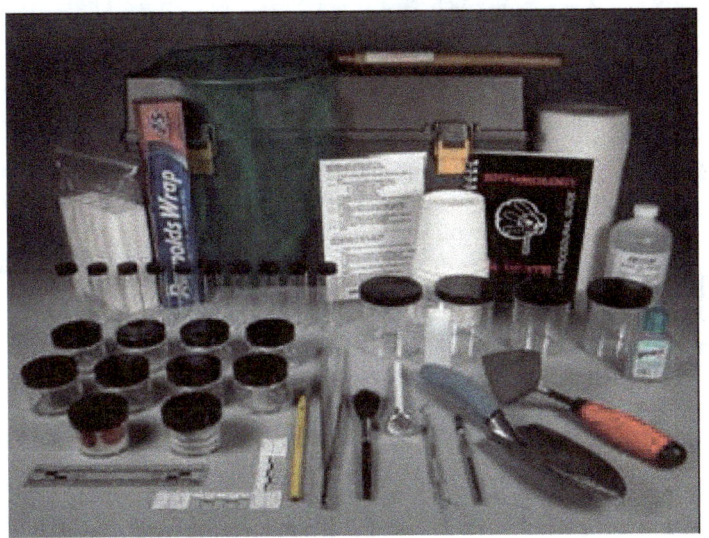

https://www.shopevident.com/category/field-kits-crime-scene

Samples of insects of all stages should be collected from different areas of the body, from the clothing and from the soil. Insects will often congregate in wounds and in and around natural orifices. The two main insect groups on bodies are flies [Diptera] and beetles [Coleoptera]. Both types of insect look very different at different stages of their lives. Flies can be found as eggs, larvae or maggots, pupae and/or empty pupal cases & adults.

Eggs - small and are usually laid in masses, and are usually found in a wound or natural orifice. They can be collected with a small paint brush dipped in water, with forceps or a bent handled spoon (seems to work best). Half should be preserved in 75% alcohol or 50% isopropyl alcohol – but make sure you put them in hot water for 30 seconds before placing them in the alcohol. The rest should be placed in a vial with a little damp tissue paper to prevent dehydration. If it will be more than a few hours before the entomologist receives them, they should also be given a small piece of beef liver.

Make sure there is tissue present if liver is added, to prevent drowning. They need some air. Newly emerged maggots can escape through holes, so a paper towel held over the top of the vial with a rubber band is excellent, as long as the vial stays upright. (No lid other than the paper towel is needed).

Maggots - collect all range of sizes. Maggots will be found crawling on or near the remains or may be in maggot masses. The masses generate a lot of heat, which speeds up development. Therefore note:

1. The site of maggot masses
2. The temperature of each mass (an accurate thermometer is necessary)
3. Label which maggots come from a particular mass location.

Large maggots are usually older so are most important, but smaller maggots may belong to a different species so both large and smaller maggots should be collected, with the emphasis on larger maggots. Collect samples of maggots from different areas of the body and the surrounding area, and keep them separate.

Third instar larvae will leave the food source to find a suitable area to pupate. They may wander some distance from the body so the soil for a meter or two around the body should be carefully sifted. Some may burrow down into leaf litter, so the soil below the corpse should be checked for several centimeters. Checking the trails away from the body will indicate larvae movement. If the remains were on a slope, the body fluids will seep downhill and insects will be found there, feeding on the fluids. This means that a very intensive search of the corpse, the clothes and the surrounding area must be made in order to get the entire picture.

When collected, a proportion of the larvae should be preserved immediately for two reasons. 1, to show the forensic entomologist (F.E.), if they're not present at the scene, what stage the larvae were in when collected, if they are placed on meat, they'll continue to develop, giving a misleading impression to the F.E. when they are examined. 2, to produce as evidence in court, if there are an abundance of maggots on the body, preserve approximately half of all sizes. If there are only 20-30, preserve 1 or 2. Preserve the specimens by immersing them in hot water for 30 seconds, then putting them in 70% alcohol or 50% isopropyl alcohol. If no hot water available, put

straight in preservative. Don't forget that most should be kept alive. A sample should contain as many as 100 maggots (of each size if possible).
The live egg specimens should be placed in a vial, with air and raw liver/beef. There should be only enough maggots to cover the bottom of the vial. Too many in one vial will drown.

Pupae and Empty Pupal Cases - these are extremely important and are easy to miss. They are often found in clothing, hair or soil near the body. Pupae like dry, secure areas away from the wet food source in which to pupate so pockets, seams and cuffs are likely hiding places. If the remains are found indoors, they may have traveled some distance and be under clothing, rugs, boxes etc. They range from 2-20 mm, and are oval, like a football. They are dark brown when completely tanned. An empty pupal case is very similar but is open at one end, where the adult fly has emerged. They need some air, so secure a paper towel over vial as for eggs, as although the pupae are immobile, if they emerge during transit, an adult can get out of anything! A piece of tissue in the vial will help to avoid breakage as they are quite vulnerable.

This needs to be moistened with water, but be careful not to drown them. The moisture isn't necessary if it's a short trip to the lab. Do not preserve pupae! They won't grow, so the reasons for preserving larvae do not apply, and it is almost impossible to identify a pupa until it emerges as an adult. The F.E. also cannot determine its exact age until they find out the day on which it emerges. If a pupa is found when a pale color, it is just entering pupation, so please keep that specimen separate and label as pale colored, as it will darken in a few hours. The specimen could be aged to a matter of hours.

Adult Flies – not as important. They are only of use in indicating which species of insect are likely to develop from the corpse, as you cannot determine whether an adult has developed on the corpse, or has just arrived from somewhere else to oviposit, unless it emerged only an hour or so earlier. If an adult has crumpled wings, it may have just emerged, so it's still important as it can be linked to the body. It should be collected, labeled as such, and kept separate. The presence of empty pupal cases, however, indicates that an insect has developed on the corpse and reached adulthood. This can be very important as it indicates that at least one generation of flies of this species has completed development on the corpse.

Beetles - can be found as adults, larvae, grubs or pupae and also as cast skins. All stages are equally important. They're often found under the body, and in and under clothing. They can be placed in vials with some air and they need to be fed if it will be more than 24 hours before they reach an F.E. They can be fed extra maggots as they are cannibals and should not be placed in the same tube/vial.
Insects - If you are not sure whether it's an insect, collect it and place in a vial. The F.E. can determine what it is later.

Samples - Soil and leaf samples will also be useful. About a 8 ounce can size of soil from under the body (first 5 cm is useful). If the soil below the body is extremely wet, it is better to collect the soil from near the remains.

Labeling - Insects collected from one part of the body should be kept separate from those from another area. Different species should be kept separate as beetle larvae feed on fly larvae. If they look different, separate them. Each vial should be labeled with:

Area of body or soil
Time and Date of collection
Person doing the collecting

Stage (larvae), so that if the specimens are pupae, the F.E. will know that they developed into the next stage during transit.
Handling - Specimens are fragile and are probably best picked up with a spoon or gloved fingers which are often gentler than forceps if you new to using them. Very tiny or delicate specimens can be picked up using an camel haired brush dipped in water or alcohol depending on what you are about to do with them. Make sure all the vials are very well sealed.

Packing - The insects should be taken to the F.E. as soon as possible. They should be couriered or hand delivered to maintain continuity. They should be packaged in a cardboard box (which is not air-tight. Each vial can be taped so that it remains upright. The whole box must remain upright.

Greg MacMaster   Environmental Forensics

Other factors about the death site needs to be documented:
Habitat -
o   Is it wooded, a beach, a house, along the roadside?
o   Trees, grass, bush, shrubs around the site
o   Type of soil - rocky, sandy, muddy
o   Weather - at time of collection, sky condition
o   Temperature and possibly humidity at collection time, wind direction, pressure.
o   GPS: Elevation and map coordinates of the death site
o   Is site in shade or direct sunlight?
o   Unusual visual clues such as whether it's possible that the body may have been submerged (partially/fully) at any time. Remains - Other "need to know" items:
o   Presence, type and amount of clothing
o   Body buried or covered in soil, cloth, leaves
o   Cause of death, if known? Is there blood at the scene?
o   Other body fluids.
o   Any visible wounds? What kind?
o   If drugs likely to be involved, this could affect the decomposition rates.
o   Position of the body (side/back/face down)
o   What direction is the body facing
o   State of decomposition
o   If a maggot mass present it could affect the temperature on the body
o   What is the temperature of the center of the maggot mass(es)? (don't probe too deep)
o   Is there any other meat or carrion around that might also attract insects?
o   Is there a possibility that death did not occur at the present site? (urban fly in country setting)

If the body is refrigerated at the morgue before the collection then the F.E. needs to know the exact time that the body went into the cooler, and the exact time it came out (noting temperature). Photographs, or a video of the scene, the body in situ and the site after removal of the body are also extremely useful. Take more pictures than necessary and utilize the zoom capability as much as possible.

When the insects reach the insectary, the immature specimens are measured, and examined, then placed in a jar containing a food. In the case of blowflies, this is usually beef liver, which is placed on top of sawdust. When the insects reach the prepupal stage and leave the food source they will burrow into the sawdust to pupate. The insects are checked daily and when they pupate they are removed and placed in a petri dish with damp filter paper. The date of pupation and the date of emergence is noted for each specimen. When the adults emerge, they are killed and pinned, then placed in an insect box. Each insect has a detailed label. Any adults collected directly from the corpse are immediately killed and pinned.

The reasons for raising the immature are two-fold. Firstly, larvae are very difficult to identify to species, but adults have many more diagnostic features. Secondly, the dates of pupation and emergence are used to help calculate the age at the time of collection.
Other important information used to determine elapsed time since death, include:-

Weather records from the nearest weather station, including temperature and precipitation

The distance between the death site and the weather station

This method of determining elapsed time since death using insect evidence can be demonstrated using an actual case. Human remains were found in mid October. Most of the head region was missing as death was due to gunshot wounds. The upper portion of the body was almost skeletonized, but the lower area, clad in tight clothes, appeared almost fresh. There were several large maggot masses on the corpse which generate their own heat for a while due to the frenzied activity. The temperature of the largest maggot mass was 20oC, even after the body had been refrigerated at 4oC for two hours. All sizes of larvae were collected and three pupae. These were pale in color so had only just pupated. No puparia were found. The mean temperature at the death site was 15C.

Two species of blowfly emerged, Calliphora vomitoria and Phormia regina. Both are common species that are amongst the first to arrive on a corpse. The oldest stage of Calliphora vomitoria collected was just entering the prepupal stage of the third instar. This was determined from size, no. of spiracular slits (breathing holes) date of pupation and behavior, in that the largest specimens immediately left the beef liver and entered the sawdust, indicating that they had stopped feeding. At the temperature of the death site, 15C, Calliphora vomitoria takes a minimum of 9.3 days to reach the beginning of the prepupal stage of the third instar. So these insects were a minimum of 9 days old when collected on 12 October, meaning that they were laid as eggs on or before 4 October. As there was blood at the scene, the insects probably arrived very soon after death. Therefore death must have occurred on or before 4 October.

Using the same techniques for Phormia regina, the oldest specimens of which were in the pupal stage when collected, it was calculated that Phormia regina was oviposited no later than 3 October. Therefore, using the two insects together, it can be shown that death occurred on or before 3 October. Other police evidence later showed that death had actually occurred on 3 October.

Post Mortem Interval by Accumulated Degree Hours/Days

Development is regarded as a combination of temperature above the minimum developmental threshold multiplied by time.

• Thus, $5°$ above the minimum developmental threshold for 2 days = 10 degree days
• And, $1°$ for 10 days = 10 degree days

- **Degree-day process involves estimating the area under a daily temperature curve that is above the minimum development threshold (generally 6° - 10°C).**

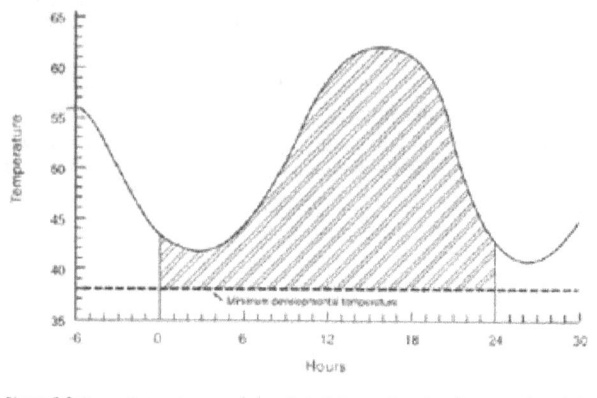

Figure 9.2 Degree-days, a measure of physiological time, indicated as the area under a daily temperature curve and above a minimum developmental temperature for a species.

- With the size of the maggot & which instar it represents known,
- Compare with data from laboratory rearings.
- Then adjust the time period to fit circumstances of the site where the body was found.
- ADH (Accumulated Degree Hours) = time
* Temperature

Sample Case to test your skill

A body is discovered at 8:00a, October 15 and insect specimens are collected & preserved at 9:00am the same day. Most mature maggot of *Phormia regina* are molting from first into second instar.

- ADH required = ADH eggs + ADH 1st instar

For laboratory rearings @ 26.7°C
- time required = 34 hours or 907.8 ADH
- Work backwards from time when maggots were collected, 9:00a, Oct. 15th
    - 9 hours of development that day with mean temp of 20°C; 9*20 = 180 ADH
    - previous day mean temp = 21°C; 21*24 = 504 ADH
    - total time for two days is 180 + 504 = 684 ADH
    - 907.8 − 684 = 223.8 ADH still required
    - if mean temp on Oct. 13th was 20°C
    - then 223.8 / 20 = 11.2 hours
    - or insect activity initiated between noon and 1pm on October 13th.

Post Mortem Interval by Days

Accumlated Degree Days (ADD)

- Accumulation of heat units above a developmental threshold temperature for an insect larvae.
- **Developmental threshold** (or thermal constant)
- is the minimum temperature necessary for the insect to feed and develop
- A developmental threshold must be known for the insect in question
- Degree Day = Average temperature - Thermal Constant (TC) of insect
- Estimate average temperature (most widely used method)=

$$DD = \frac{Max\ temp + Min\ Temp}{2} - TC$$

Example of a Green Bottle Blow fly

TC = 7 degrees C
    If max. temp = 36 and min = 30 then = 36 + 30 =
    66 / 2 = 33 average temp
    33 - 7 = 26 degree day
In practice, ADD is worked backwards.

Larvae are collected at the scene and reared to adulthood
> If it took 6 days in the lab to reach the adult stage at 30 degree days per day and the life cycle developmental data says it takes 300 degree days to reach adulthood

Then 300 - 180 = 120 degree days old when collected.
Then looking at the outside temperature records you can determine how many degree days were possible and pinpoint time of oviposition.

Many common Minimum Developmental Thresholds have been accumulated in (CRC book Forensic Entomology edited by J.H. Byrd and J.L. Castner)

Of course there are influences on the development and succession which are described below.

- Geography
- Seasonal (defined later in this chapter)
- Exposure to sun (even in partial shade)
- Urban vs. rural (City vs. Country)
- Bodies found inside buildings
- Buried bodies
- Bodies in water
- Bodies in vehicles
- Bodies in enclosed spaces
- Hanged bodies
- Burnt remains
- Wrapped remains
- Drugs and other toxins
- Food Type

OTHER USES FOR INSECTS IN FORENSIC SCIENCE

The body may have been moved after death, from the scene of the killing to a hiding place. Some of the insects on the body may be native to the first habitat and not the second. This will show that not only was the body moved, but it will also give an indication of the type of area where the murder actually took place.

The body may have been disturbed after death, by the killer returning to the scene of the crime. This may disturb the insects cycle, and the entomologist may be able to determine not only the date of death, but also the date of the return of the killer.

The presence and position of wounds, decomposition may obscure wounds. Insects colonize remains in a specific pattern, usually laying eggs first in the facial orifices, unless there are wounds, in which case they will colonize these first, then proceed down the body. If the maggot activity is centered away from the natural orifices, then it is likely that this is the site of a wound. For example, maggot activity on the palm of the hands indicates the probable presence of defense wounds.

The presence of drugs can be determined using insect evidence. There is often not enough flesh left to determine drug presence, but maggots bioaccumulate and can be analyzed to determine type of drug present.

Insects can be used to place a suspect at the scene of a crime. For instance, a grasshopper legs was found in the crease of a man trousers which led authorities to believe he was a prime suspect. The death of the woman had the other parts of the grasshopper on her. Civil cases also sometimes use insect evidence.

Child or senior abuse/neglect: Some insects will colonize around wounds or unclean areas on a living person. This is called cutaneous myiasis. If the victim is still alive, but maggot infested, a forensic entomologist will be able to tell when the wound or abuse occurred. If there's a case of a neglected child, the onset of maggot infestation will give a minimum time interval since the child last had a diaper change. Such cases occur particularly in young children and seniors. Although forensic entomology can be very effective in determining elapsed time since death, it has its limitations:-

The **temperature** of the death site is obviously a very important factor, but few criminals are thoughtful enough to kill their victim right underneath a weather station! In most cases, the weather records come from several miles away. We are trying to overcome this by setting up a miniature weather station at the death site after discovery, to compare these data with that from the weather station, in order to determine the difference between the two sites, if any.

Also the microclimate of the corpse itself will be slightly different from the surrounding area, especially if a maggot mass is present. Therefore, it is extremely important to know whether masses are present.

Studies have been performed to correlate the weather at the crime scene and closest official weather station. The best way to document the difference between the two sites is to maintain an hourly record of weather elements pertinent to the case (Temperature, Wind direction and speed). Solar radiation and cloud cover can be obtained from records and satellite images can verify the sky condition.

After a period of 24-48 hours of recording, compare with the official records and then make adjustments to determine best probable dataset for use in rearing insects.

Forensic entomology is **seasonal**, that is, it is only commonly used in spring, summer, and fall when insects are abundant. It is of less use in winter, unless it's very mild, as there are no or very few insects present. This can be a limitation, but can also be an advantage as I can sometimes show that a victim found in spring was killed the previous fall if insect evidence is present.

The results are not immediate, as it takes time to rear the insects. DNA evidence is now being developed to speed up identification of immature specimens.

The body may have been disposed of in a way that **excludes** insects' *e.g.*

Freezing - if the body was frozen for a period of time before being placed outside on, for example, 8 May, the insects would only invade then, giving the misleading impression that death had occurred on 8 May. However, other forensic experts would be able to determine whether or not the body has been frozen, and insect evidence will still determine time of exposure.

Burial - if the body is buried deeply, then most insects will be excluded. However, most criminal burials are not very deep, as the aim is merely to conceal the body, and most insects will dig down to the body, particularly if there is blood soaked in the soil. Therefore, insect evidence can still be used.

Wrapped - if the body is wrapped or packaged in some way the insects may be excluded, but the wrapping must be completely secure. A body part was found sealed in a garbage bag which had been tied securely at the top, but the remains were maggot-infested, and showed severe insect damage. The adult females had probably laid their eggs at the knot, and the minute first instar larvae had crawled in.

More **research** is needed. Insect succession varies from geographic region to region and the species and time of colonization must be developed for all areas using this type of evidence. Research has been conducted in British Columbia in a variety of habitats, seasons and geographic areas to develop a database for this Province. It is intended that this will be extended across Canada.

Drugs/Insecticides - the presence of this may affect the development of the insects and, although there may not be a direct correlation as it applies to the atmosphere, a better definition of toxicology follows in this chapter.

In conclusion, **INSECTS ARE EVIDENCE!** Forensic entomology is a very useful method of determining elapsed time since death after 72 h. It is accurate to a day or less, or a range of days, and may be the only method available to determine elapsed time since death. It is vital that the insects are collected properly and its accuracy depends on this and on suitable conditions for insects. Knowing the effects temperature has on the rearings will help the forensic meteorologist in their research and provide more accurate information to the FE.

Questions for Discussions

1. Find, from an entomologist, the process they use in the growth rate of insects.

2. If a constant temperature is used in a controlled environment, would the growth rate change if the temperature was adjusted to match the outside temperature? What would the difference be in hours?

3. Which insects visit the decaying body first? Why?

4. If insects can be used to place a suspect at the scene of a crime, can the change in temperature alter the time-line of the growth rate of insects and change the time-line of other possible suspects? Discuss.

Chapter 6
FORENSIC TOXICOLOGY

Forensic toxicology is essentially a specialty area of analytical chemistry. *Toxicology* is the science of adverse effects of chemicals on living organisms. In general, a toxicologist detects and identifies foreign chemicals in the body, with a particular emphasis upon toxic or hazardous substances. A *descriptive toxicologist* performs toxicity tests to evaluate the risk that exposure poses to humans. A *mechanistic toxicologist* attempts to determine how substances exert deleterious effects on living organisms. A *regulatory toxicologist* judges whether or not a substance has low enough risk to justify making it available to the public.

A *toxin* is any material exerting a life threatening effect upon a living organism. Poisons are a subgroup of toxins. Toxic materials exist in many forms (gaseous, liquid, solid, animal, mineral, and vegetable), and may be ingested, inhaled, or absorbed through the skin. *Poisons* generally enter the body in a single massive dose, or accumulate to a massive dose over time. Toxins work in minute quantities or low levels, requiring sensitive analytical instruments for detection. Some toxins have medicinal value, but many produce irreparable damage. Some toxins have antidotes and others do not. Poisons can be combated by prompt treatment, and most organ damage (except for serious CNS injury) may be repairable. Whereas poisons are somewhat easily identifiable by their symptoms, many toxins tend to disguise or mask themselves.

The toxic effects of substances are not side effects. Side effects are defined as non-deleterious, such as dry mouth, for example. Toxic effects are the undesirable results of a direct effect. They occur in a number of ways, most often produced by a dangerous metabolite of the drug which is activated by an enzyme, light, or oxygen reaction in a process known as *biotransformation*. Toxic reactions often depend on how metabolites are processed by an individual's body, how proteins build up and bind at effector sites in the body. Some metabolites destroy liver cells, others brain tissue, and still others operate at the DNA level. Toxic reactions are classified as one of three (3) reactions:

- pharmacological -- injury to the central nervous system (CNS)
- pathological -- injury to the liver
- genotoxic -- creation of benign or malignant neoplasms or tumors

If the concentration of toxin doesn't reach a *critical level*, the effects will usually be reversible. Pharmacological reactions, for example, are of this type. In order to sustain permanent brain damage, dosages must be above a standard critical level. Pathological reactions can be repaired if discovered early enough, but most liver damage occurs over a period of few months to a decade. Genotoxic or carcinogenic effects may take 20-40 years before tumors develop. Most of the time, toxic metabolites are activated by enzymatic transformation, but a few are activated by light. This means that exposure of the skin to sunlight produces a photoallergic reaction or *phototoxic reaction* within 24 hours. It's important to understand that *the target organ of toxicity is not the site where toxin accumulates.*

Lead poisoning, for example, results in an accumulation of lead in bone marrow, but the toxic effect is the creation of lesions on skin and soft tissue. Carcinogenesis is even more complicated, involving the creation of promotor electrophiles which serve to activate or potentiate the growth of latent tumors given some biological trigger or subsequent environmental attack. Different people, of course, have *chemical allergies* (as well as food allergies), depending upon the serology of their allergen-antigen history. In such people, toxic reactions take different forms. Other people have what are called *idosyncratic reactions*, which means they have certain unique genetic triggers. Furthermore, people exposed to multiple toxins can have *synergistic reactions*, which means that two or more toxins interact at the metabolic level to be greater or less than the effects of the individual toxins.

A forensic toxicologist is normally presented with preserved samples of body fluids, stomach contents, and organ parts. They will have access to the coroner's report which should contain information on various signs and symptoms as well as other postmortem data. The toxicologist needs a through knowledge of how the body alters or metabolizes drugs because few substances leave the body in the same state as they entered. The substances they work with are often *derivatives*, which is a term meaning a chemical compound which is prepared from a pure compound in order to be more easily detected by the analytical techniques used. They also divide specimens up into acidic and basic *fractions* for drug extraction from tissue or fluid. Almost all drugs are either acids or bases (on a pH scale from 0 to 14 with closer to 0 being acids and closer to 14 being bases). Acid drugs are easily extracted with a pH solution of less than 7; base drugs are easily extracted with a pH solution of greater than 7. As an example, most of the barbiturate drugs are acid-soluble; most of the amphetamine drugs are base-soluble.

After preliminary acid-base procedures are carried out, and the tissue or fluid sample is now a drug sample, examination continues in two steps: (1) screening tests, and (2) confirmation testing. Screening tests allow the processing of many specimens for a wide range of toxins in a short time. Any positive indications from the screening tests must be verified with a confirmation test. The following are some standard laboratory tests for toxin detection:

SCREENING TESTS

- Physical tests -- boiling point, melting point, density, and refractive index

- Crystal tests -- treatment with a chemical reagent to produce crystals

- Chemical spot tests -- treatment with a chemical reagent to produce color changes

- Chromatography (thin-layer or gas) -- used to separate components of a mixture

CONFIRMATION TESTS

- Mass spectrometry -- this is a combination of gas chromatography/mass spectrometry which is generally accepted as the confirmation test of choice. Each toxin has a known mass spectra, or "fingerprint", which is infallible proof of its presence at the chemical level

Ordinarily, the toxicologist is not required to render an opinion of whether the toxin levels in the body were enough to cause death. A few toxicologists may do so, but they must have had special training in physiology, and this is usually the province of the forensic pathologist, in any event. Often, the defense will call their medical experts to dispute a cause of death claim. Physicians are the only ones qualified to render opinions on the physiological effects of toxins, and forensic law allows them to provide their testimony in the form of *hypotheticals*, even though they do not have personal knowledge of the case. Low-level toxin cases usually become a real battle of the experts.

Drug overdoses and alcoholic poisonings will provide most of the work for toxicologists; hence a couple of allied subfields may be drawn upon: (1) a field inhabited by what are called Drug Recognition Experts (DRE); and (2) alcohol intoxication measurement (a subject talked about in a previous lecture). Both are sought-after areas of police training. Another related subfield involves carbon monoxide poisoning, which may involve an automobile engineer or fire safety specialist.

The Drug Recognition Expertise evolved out of experiments in California with the LAPD during the 1970s in which police officers were trained to identify and recognize certain types of drugs based upon the impairments and physiological symptoms. The examination that such specially trained police officers conduct goes beyond normal Nystagmus testing and more closely resembles the taking of vital signs by a nurse or paramedic, combined with structured interviewing and observation.

Lastly, there will be times when a railcar, or tanker, will derail and a chemical spill will occur. A hazardous response team should already be in place but just in-case, take the time to obtain current weather and immediate forecast that will impact areas downwind. There are computer programs that will calculate this based on the chemical and weather conditions, but they are expensive. Ask a member of the local hazardous response team to visit the class and discuss toxic corridor applications or visit their place of business and see how they use the 'weather' in their line of work.

Benjamin, D. (1993). "Forensic Pharmacology" in R. Saferstein (ed.) *Forensic Science Handbook*. NJ: Prentice-Hall.

Klaasen, C. (1996). "Principles of Toxicology and Treatment of Poisoning" in J. Hardman et al., *Goodman and Gilman's The Pharmacological Basis of Therapeutics*. NY: McGraw-Hill.

Levine, A. (1993). "Forensic Toxicology" *Journal of Analytical Chemistry* 65: 272-76.

Lowry, W. & J. Garriott. (1979). *Forensic Toxicology: Controlled Substances and Dangerous Drugs*. NY: Plenum.

Moenssens, A.A.; Inbau, F.E.; Starrs, J.E. (1986). Scientific Evidence in Civil and Criminal Cases. NY: The Foundation Press.

Saferstein, R. (1998). *Criminalistics: An Introduction to Forensic Science*. NJ: Prentice-Hall.

Questions for Discussion

1. While toxicology testing may not be dependent on temperature, can you define an instance where it would be dependent on any weather element?

2. An outbreak occurs when legions are found on many people who are isolated in a specific geographic location within or near a city. You are asked to verify wind speed and direction from a certain time-frame. Why is this data requested?

3. How would winds play a roll in an airborne chemical outbreak?

4. What weather elements affect windborne contamination from a chemical spill? Discuss with class.

Chapter 7

NATIONAL GUIDELINES FOR DEATH INVESTIGATION

In June and July 1997, NMRP met for two 11/2-day working sessions in St. Louis and Chicago to review the draft guidelines developed by the Executive board and offer recommendations and changes based on Jurisdictional variances and organizational responsibilities. Those sessions resulted in the final draft of the 29 guidelines for conducting death investigations. The 29 guidelines are presented in the next main section. Notice the lack of weather recognition in terms of observing the elements of weather in various forms. **Having knowledge of the investigation process can help you determine how best to provide weather information that will fit their needs better.**

Guideline Status

Currently, NMRP members are presenting the guidelines to their respective organizations' leadership (or appropriate internal committees) for review. This researcher is collecting anecdotal comments for future modification of the existing guidelines during the validation procedures.

Training Guidelines

The purpose of the second part of the national death investigator guidelines research was to identify training criteria for each of the 29 guidelines. This research is now completed. (A death investigation curriculum guide and a CD-ROM for use in the field or classroom are in development.) For each of the guidelines presented in this report, "minimum levels of performance" will be developed and verified by the members of the various TWGs. These "training guidelines" will provide both individuals and educational organizations the material needed to establish and maintain valid exit outcomes for each investigative trainee.

Guideline Validation

In this initial research, 29 investigative tasks were identified. Each task was developed into a guideline for investigators to follow while conducting a death investigation. Although each TWG believed in the validity of each guideline, no attempt was made to validate actual significance (e.g., if guideline C1 is trained and implemented, a [%] decrease in poor scene photographs should occur). The researcher is currently developing a national validation strategy for the implementation and validation of each guideline.

This handbook is intended as a guide to recommended practices for the investigation of death scenes. Jurisdictional, logistical, or legal conditions may preclude the use of particular procedures contained herein.

Section A: Investigative Tools and Equipment

1. Gloves (Universal Precautions).
2. Writing implements (pens, pencils, markers).
3. Body bags.
4. Communication equipment (cell phone, pager, radio).
5. Flashlight.
6. Body ID tags.
7. Camera--35mm (with extra batteries, film, etc.).
8. Investigative notebook (for scene notes, etc.).
9. Measurement instruments (tape measure, ruler, rolling measuring tape, etc.).
10. Official identification (for yourself).
11. Watch.
12. Paper bags (for hands, feet, etc.).
13. Specimen containers (for evidence items and toxicology specimens).
14. Disinfectant (Universal Precautions).
15. Departmental scene forms.
16. Camera--Polaroid (with extra film).
17. Blood collection tubes (syringes and needles).
18. Inventory lists (clothes, drugs, etc.).
19. Paper envelopes.
20. Clean white linen sheet (stored in plastic bag).
21. Evidence tape.

22. Business cards/office cards w/phone numbers.
23. Foul-weather gear (raincoat, umbrella, etc.).
24. Medical equipment kit (scissors, forceps, tweezers, exposure suit, scalpel handle, blades, disposable syringe, large gauge needles, cotton-tipped swabs, etc.).
25. Phone listing (important phone numbers).
26. Tape or rubber bands.
27. Disposable (paper) jumpsuits, hair covers, face shield, etc.
28. Evidence seal (use with body bags/locks).
29. Pocketknife.
30. Shoe-covers.
31. Trace evidence kit (tape, etc.).
32. Waterless hand wash.
33. Thermometer.
34. Crime scene tape.
35. First aid kit.
36. Latent print kit.
37. Local maps.
38. Plastic trash bags.
39. Gunshot residue analysis kits (SEM/EDS).
40. Photo placards (signage to ID case in photo).
41. Boots (for wet conditions, construction sites, etc.).
42. Hand lens (magnifying glass).
43. Portable electric area lighting.
44. Barrier sheeting (to shield body/area from public view).
45. Purification mask (disposable).
46. Reflective vest.
47. Tape recorder.
48. Basic handtools (boltcutter, screwdrivers, hammer, shovel, trowel, paintbrushes, etc.).
49. Body bag locks (to secure body inside bag).
50. Camera--Video (with extra battery).
51. Personal comfort supplies (insect spray, sun screen, hat, etc.).
52. Presumptive blood test kit.

Section B: Arriving at the Scene

1. Introduce and Identify Self and Role

Principle: Introductions at the scene allow the investigator to establish formal contact with other official agency representatives. The investigator must identify the first responder to ascertain if any artifacts or contamination may have been introduced to the death scene. The investigator must work with all key people to ensure scene safety prior to his/her entrance into the scene.

Authorization: Medical Examiner/Coroner Official Office Policy Manual;
State or Federal Statutory Authority.

Policy: The investigator shall take the initiative to introduce himself or herself, identify essential personnel, establish rapport, and determine scene safety.

Procedure: Upon arrival at the scene, and prior to entering the scene, the investigator should:

A. Identify the lead investigator at the scene and present identification.
B. Identify other essential officials at the scene (e.g., law enforcement, fire, EMS, social/child protective services, etc.) and explain the investigator's role in the investigation.
C. Identify and document the identity of the first essential official(s) to the scene (first "professional" arrival at the scene for investigative follow up) to ascertain if any artifacts or contamination may have been introduced to the death scene.
D. Determine the scene safety (prior to entry).

Summary: Introductions at the scene help to establish a collaborative investigative effort. It is essential to carry identification in the event of questioned authority. It is essential to establish scene safety prior to entry.

2. Exercise Scene Safety

Principle: Determining scene safety for all investigative personnel is essential to the investigative process. The risk of environmental and physical injury must be removed prior to initiating a scene investigation. Risks can include hostile crowds, collapsing structures, traffic, and environmental and chemical threats.

Authorization: Medical Examiner/Coroner Official Office Policy Manual;
State or Federal Statutory Authority.

Policy: The investigator shall attempt to establish scene safety prior to entering the scene to prevent injury or loss of life, including contacting appropriate agencies for assistance with other scene safety issues.

Procedure: Upon arrival at the scene, the investigator should:

A. Assess and/or establish physical boundaries.
B. Identify incident command.
C. Secure vehicle and park as safely as possible.
D. Use personal protective safety devices (physical, biochemical safety).
E. Arrange for removal of animals or secure (if present and possible).
F. Obtain clearance/authorization to enter scene from the individual responsible for scene safety (e.g., fire marshal, disaster coordinator).
G. While exercising scene safety, protect the integrity of the scene and evidence to the extent possible from contamination or loss by people, animals, and elements.

Note: Due to potential scene hazards (e.g., crowd control, collapsing structures, poisonous gases, traffic), the body may have to be removed before scene investigation can be continued.

Summary:

Environmental and physical threats to the investigator must be removed in order to conduct a scene investigation safely. Protective devices must be used by investigative staff to prevent injury. The investigator must endeavor to protect the evidence against contamination or loss.

3. Confirm or Pronounce Death

Principle: Appropriate personnel must make a determination of death prior to the initiation of the death investigation. The confirmation or pronouncement of death determines jurisdictional responsibilities.

Authorization: Medical Examiner/Coroner Official Office Policy Manual; State or Federal Statutory Authority.

Policy: The investigator shall ensure that appropriate personnel have viewed the body and that death has been confirmed.

Procedure: Upon arrival at the scene, the investigator should:

A. Locate and view the body.
B. Check for pulse, respiration, and reflexes, as appropriate.
C. Identify and document the individual who made the official determination of death, including the date and time of determination.
D. Ensure death is pronounced, as required.

Summary:

Once death has been determined, rescue/resuscitative efforts cease and medico-legal jurisdiction can be established. It is vital that this occur prior to the medical examiner/coroner's assuming any responsibilities.

4. Participate in Scene Briefing (With Attending Agency Representatives)

Principle: Scene investigators must recognize the varying jurisdictional and statutory responsibilities that apply to individual agency representatives (e.g., law enforcement, fire, EMT, judicial/legal). Determining each agency's investigative responsibility at the scene is essential in planning the scope and depth of each scene investigation and the release of information to the public.

Authorization: Medical Examiner/Coroner Official Office Policy Manual; State or Federal Statutory Authority.

Policy: The investigator shall identify specific responsibilities, share appropriate preliminary information, and establish investigative goals of each agency present at the scene.

Procedure: When participating in scene briefing, the investigator should:

A. Locate the staging area (entry point to scene, command post, etc.).
B. Document the scene location (address, mile marker, building name) consistent with other agencies.
C. Determine nature and scope of investigation by obtaining preliminary investigative details (e.g., suspicious versus nonsuspicious death).
D. Ensure that initial accounts of incident are obtained from the first witness(es).

Summary:

Scene briefing allows for initial and factual information exchange. This includes scene location, time factors, initial witness information, agency responsibilities, and investigative strategy.

5. Conduct Scene "Walk Through"

Principle: Conducting a scene "walk through" provides the investigator with an overview of the entire scene. The "walk through" provides the investigator with the first opportunity to locate and view the body, identify valuable and/or fragile evidence, and determine initial investigative procedures providing for a systematic examination and documentation of
the scene and body.

Authorization: Medical Examiner/Coroner Official Office Policy Manual; State or Federal Statutory Authority.

Policy: The investigator shall conduct a scene "walk through" to establish pertinent scene parameters.

Procedure: Upon arrival at the scene, the investigator should:

A. Reassess scene boundaries and adjust as appropriate.
B. Establish a path of entry and exit.
C. Identify visible physical and fragile evidence.
D. Document and photograph fragile evidence immediately and collect if appropriate.
E. Locate and view the decedent.

Summary:

The initial scene "walk through" is essential to minimize scene disturbance and to prevent the loss and/or contamination of physical and fragile evidence.

6. Establish Chain of Custody

Principle: Ensuring the integrity of the evidence by establishing and maintaining a chain of custody is vital to an investigation. This will safeguard against subsequent allegations of tampering, theft, planting, and contamination of evidence.

Authorization: Medical Examiner/Coroner Official Office Policy Manual; State or Federal Statutory Authority.

Policy: Prior to the removal of any evidence, the custodian(s) of evidence shall be designated and shall generate and maintain a chain of custody for all evidence collected.

Procedure: Throughout the investigation, those responsible for preserving the chain of custody should:

A. Document location of the scene and time of arrival of the death investigator at the scene.
B. Determine custodian(s) of evidence, determine which agency(ies) is/are responsible for collection of specific types of evidence, and determine evidence collection priority for fragile/fleeting evidence.
C. Identify, secure, and preserve evidence with proper containers, labels, and preservatives.

This is the time when a four sample reading of a rectal temperature would be needed. (4 - 10 minute readings logging body temp, air temp, ground temp and wind speed). This information would be beneficial when entering the data into the new ETOD computer program that is currently in development.

D. Document the collection of evidence by recording its location at the scene, time of collection, and time and location of disposition.
E. Develop personnel lists, witness lists, and documentation of times of arrival and departure of personnel.

Summary:

It is essential to maintain a proper chain of custody for evidence. Through proper documentation, collection, and preservation, the integrity of the evidence can be assured. A properly maintained chain of custody and prompt transfer will reduce the likelihood of a challenge to the integrity of the evidence.

7. Follow Laws (Related to the Collection of Evidence)

Principle: The investigator must follow local, State, and Federal laws for the collection of evidence to ensure its admissibility. The investigator must work with law enforcement and the legal authorities to determine laws regarding collection of evidence.

Authorization: Medical Examiner/Coroner Official Office Policy Manual; State or Federal Statutory Authority.

Policy: The investigator working with other agencies must identify and work under appropriate legal authority. Modification of informal procedures may be necessary but laws must always be followed.

Procedure: The investigator, prior to or upon arrival at the death scene, should work with other agencies to:

A. Determine the need for a search warrant (discuss with appropriate agencies).
B. Identify local, State, Federal, and international laws (discuss with appropriate agencies).
C. Identify medical examiner/coroner statutes and/or office standard operating procedures (discuss with appropriate agencies).

Summary:

Following laws related to the collection of evidence will ensure a complete and proper investigation in compliance with State and local laws, admissibility in court, and adherence to office policies and protocols.

Section C: Documenting and Evaluating the Scene

Greg MacMaster   Environmental Forensics

1. Photograph Scene

Principle: The photographic documentation of the scene creates a permanent historical record of the scene. Photographs provide detailed corroborating evidence that constructs a system of redundancy should questions arise concerning the report, witness statements, or position of evidence at the scene.

Authorization: Medical Examiner/Coroner Official Office Policy Manual; State or Federal Statutory Authority.

Policy: The investigator shall obtain detailed photographic documentation of the scene that provides both instant and permanent high-quality (e.g., 35 mm) images.

Procedure: Upon arrival at the scene, and prior to moving the body or evidence, the investigator should:

A. Remove all nonessential personnel from the scene.
B. Obtain an overall (wide-angle) view of the scene to spatially locate the specific scene to the surrounding area.
C. Photograph specific areas of the scene to provide more detailed views of specific areas within the larger scene.
D. Photograph the scene from different angles to provide various perspectives that may uncover additional evidence.
E. Obtain some photographs with scales to document specific evidence.
F. Obtain photographs even if the body or other evidence has been moved.

Note: If evidence has been moved prior to photography, it should be noted in the report, but the body or other evidence should not be reintroduced into the scene in order to take photographs.

Summary:

Photography allows for the best permanent documentation of the death scene. It is essential that accurate scene photographs are available for other investigators, agencies, and authorities to recreate the scene. Photographs are a permanent record of the terminal event and retain evidentiary value and authenticity. It is essential that the investigator obtain accurate photographs before releasing the scene.

2. Develop Descriptive Documentation of the Scene

Principle: Written documentation of the scene(s) provides a permanent record that may be used to correlate with and enhance photographic documentation, refresh recollections, and record observations.

Authorization: Medical Examiner/Coroner Official Office Policy Manual; State or Federal Statutory Authority.

Policy: Investigators shall provide written scene documentation.

Procedure: After photographic documentation of the scene and prior to removal of the body or other evidence, the investigator should:

A. Diagram/describe in writing items of evidence and their relationship to the body with necessary measurements.
B. Describe and document, with necessary measurements, blood and body fluid evidence including volume, patterns, spatters, and other characteristics.
C. Describe scene environments including odors, lights, temperatures, and other fragile evidence.

Note: If evidence has been moved prior to written documentation, it should be noted in the report.

Summary:

Written scene documentation is essential to correlate with photographic
evidence and to recreate the scene for police, forensic(s), and judicial and civil agencies with a legitimate interest.

3. Establish Probable Location of Injury or Illness

Principle: The location where the decedent is found may not be the actual location where the injury/illness that contributed to the death occurred. It is imperative that the investigator attempt to determine

the locations of any and all injury(ies)/illness(es) that may have contributed to the death.
Physical evidence at any and all locations may be pertinent in establishing the cause, manner, and circumstances of death.

Authorization: Medical Examiner/Coroner Official Office Policy Manual; State or Federal Statutory Authority.

Policy: The investigator shall obtain detailed information regarding any and all probable locations associated with the individual's death.

Procedure: The investigator should:

A. Document location where death was confirmed.
B. Determine location from which decedent was transported and how body was transported to scene.
C. Identify and record discrepancies in rigor mortis, livor mortis, and body temperature.
D. Check body, clothing, and scene for consistency/inconsistency of trace
evidence and indicate location where artifacts are found.
E. Check for drag marks (on body and ground).
F. Establish post-injury activity.
G. Obtain dispatch (e.g., police, ambulance) record(s).
H. Interview family members and associates as needed.

Summary:

Due to post-injury survival, advances in emergency medical services, multiple modes of transportation, the availability of specialized care, or criminal activity, a body may be moved from the actual location of illness/injury to a remote site. It is imperative that the investigator attempt to determine any and all locations where the decedent has previously been and the mode of transport from these sites.

4. Collect, Inventory, and Safeguard Property and Evidence

Principle: The decedent's valuables/property must be safeguarded to ensure proper processing and eventual return to next of kin. Evidence on or near the body must be safeguarded to ensure its availability for further evaluation.

Authorization: Medical Examiner/Coroner Official Office Policy Manual; State or Federal Statutory Authority.

Policy: The investigator shall ensure that all property and evidence is collected, inventoried, safeguarded, and released as required by law.

Procedure: After personal property and evidence have been identified at the scene, the investigator (with a witness) should:

A. Inventory, collect, and safeguard illicit drugs and paraphernalia at scene and/or office.
B. Inventory, collect, and safeguard prescription medication at scene and/or office.
C. Inventory, collect, and safeguard over-the-counter medications at scene and/or office.
D. Inventory, collect, and safeguard money at scene and at office.
E. Inventory, collect, and safeguard personal valuables/property at scene and at office.

Summary:

Personal property and evidence are important items at a death investigation. Evidence must be safeguarded to ensure its availability if needed for future evaluation and litigation. Personal property must be safeguarded to ensure its eventual distribution to appropriate agencies or individuals and to reduce the likelihood that the investigator will be accused of stealing property.

5. Interview Witness(es) at the Scene

Principle: The documented comments of witnesses at the scene allow the investigator to obtain primary source data regarding discovery of body, witness corroboration, and terminal history. The documented interview provides essential information for the investigative process.

Authorization: Medical Examiner/Coroner Official Office Policy Manual; State or Federal Statutory Authority.

Policy: The investigator's report shall include the source of information, including specific statements and information provided by the witness.

Procedure: Upon arriving at the scene, the investigator should:

A. Collect all available identifying data on witnesses (e.g., full name, address, DOB, work and home telephone numbers, etc.).
B. Establish witness' relationship/association to the deceased.
C. Establish the basis of witness' knowledge (how does witness have knowledge of the death?).
D. Obtain information from each witness.

E. Note discrepancies from the scene briefing (challenge, explain, verify statements).
F. Tape statements where such equipment is available and retain them.

Summary:

The final report must document witness' identity and must include a summary of witness' statements, corroboration with other witnesses, and the circumstances of discovery of the death. This documentation must exist as a permanent record to establish a chain of events.

Section D: Documenting and Evaluating the Body

1. Photograph the Body

Principle: The photographic documentation of the body at the scene creates a permanent record that preserves essential details of the body position, appearance, identity, and final movements. Photographs allow sharing of information with other agencies investigating the death.

Authorization: Medical Examiner/Coroner Official Office Policy Manual; State or Federal Statutory Authority.

Policy: The investigator shall obtain detailed photographic documentation of the body that provides both instant and permanent high-quality (e.g., 35 mm) images.

Procedure: Upon arrival at the scene, and prior to moving the body or evidence, the investigator should:

A. Photograph the body and immediate scene (including the decedent as initially found).
B. Photograph the decedent's face.
C. Take additional photographs after removal of objects/items that interfere with photographic documentation of the decedent (e.g., body removed from car).
D. Photograph the decedent with and without measurements (as appropriate).
E. Photograph the surface beneath the body (after the body has been removed, as appropriate).

Note: Never clean face, do not change condition. Take multiple shots if possible.

Summary:

The photographic documentation of the body at the scene provides for documentation of the body position, identity, and appearance. The details of the body at the scene provide investigators with pertinent information of the terminal events.

2. Conduct External Body Examination (Superficial)

Principle: Conducting the external body examination provides the investigator with objective data regarding the single most important piece of evidence at the scene, the body. This documentation provides detailed information regarding the decedent's physical attributes, his/her relationship to the scene, and possible cause, manner, and circumstances of death.

Authorization: Medical Examiner/Coroner Official Office Policy Manual; State or Federal Statutory Authority.

Policy: The investigator shall obtain detailed photographs and written documentation of the decedent at the scene.

Procedure: After arrival at the scene and prior to moving the decedent, the investigator should, without removing decedent's clothing:

A. Photograph the scene, including the decedent as initially found and the surface beneath the body after the body has been removed.

Note: If necessary, take additional photographs after removal of objects/items that interfere with photographic documentation of the decedent.

B. Photograph the decedent with and without measurements (as appropriate), including a photograph of the decedent's face.
C. Document the decedent's position with and without measurements (as appropriate).
D. Document the decedent's physical characteristics.
E. Document the presence or absence of clothing and personal effects.
F. Document the presence or absence of any items/objects that may be relevant.
G. Document the presence or absence of marks, scars, and tattoos.
H. Document the presence or absence of injury/trauma, petechiae, etc.
I. Document the presence of treatment or resuscitative efforts.
J. Based on the findings, determine the need for further evaluation/assistance of forensic specialists (e.g., pathologists, odontologists).

Summary:

Thorough evaluation and documentation (photographic and written) of the deceased at the scene is essential to determine the depth and direction the investigation will take.

3. Preserve Evidence (on Body)

Principle: The photographic and written documentation of evidence on the body allows the investigator to obtain a permanent historical record of that evidence. To maintain chain of custody, evidence must be collected, preserved, and transported properly. In addition to all of the physical

evidence visible on the body, blood and other body fluids present must be photographed and documented prior to collection and transport. Fragile evidence (that which can be easily contaminated, lost, or altered) must also be collected and/or preserved to maintain chain of custody and to assist in
determination of cause, manner, and circumstances of death.

Authorization: Medical Examiner/Coroner Official Office Policy Manual; State or Federal Statutory Authority.

Policy: With photographic and written documentation, the investigator will provide a permanent record of evidence that is on the body.

Procedure: Once evidence on the body is recognized, the investigator should:

A. Photograph the evidence.
B. Document blood/body fluid on the body (froth/purge, substances from orifices), location, and pattern before transporting.
C. Place decedent's hands and/or feet in unused paper bags (as determined by the scene).
D. Collect trace evidence before transporting the body (e.g., blood, hair, fibers, etc.).
E. Arrange for the collection and transport of evidence at the scene (when necessary).
F. Ensure the proper collection of blood and body fluids for subsequent analysis (if body will be released from scene to an outside agency without an autopsy).

Summary:

It is essential that evidence be collected, preserved, transported, and documented in an orderly and proper fashion to ensure the chain of custody and admissibility in a legal action. The preservation and documentation of the evidence on the body must be initiated by the investigator at the scene to prevent alterations or contamination.

4. Establish Decedent Identification

Principle: The establishment or confirmation of the decedent's identity is paramount to the death investigation. Proper identification allows notification of next of kin, settlement of estates, resolution of criminal and civil litigation, and the proper completion of the death certificate.

Authorization: Medical Examiner/Coroner Official Office Policy Manual; State or Federal Statutory Authority.

Policy: The investigator shall engage in a diligent effort to establish/confirm the decedent's identity.

Procedure: To establish identity, the investigator should document use of the following methods:

A. Direct visual or photographic identification of the decedent if visually recognizable.
B. Scientific methods such as fingerprints, dental, radiographic, and DNA comparisons.
C. Circumstantial methods such as (but not restricted to) personal effects, circumstances, physical characteristics, tattoos, and anthropologic data.

Summary:

There are several methods available that can be used to properly identify deceased persons. This is essential for investigative, judicial, family, and vital records issues.

5. Document Post Mortem Changes

Principle: The documenting of post mortem changes to the body assists the investigator in explaining body appearance in the interval following death. Inconsistencies between post mortem changes and body location may indicate movement of body and validate or invalidate witness statements. In addition, post mortem changes to the body, when correlated with circumstantial information, can assist the investigators in estimating the approximate time of death.

Authorization: Medical Examiner/Coroner Official Office Policy Manual; State or Federal Statutory Authority.

Policy: The investigator shall document all post mortem changes relative to the decedent and the environment.

Procedure: Upon arrival at the scene and prior to moving the body, the investigator should note the presence of each of the following in his/her report:

A. Livor (color, location, blanchability, Tardieu spots) consistent/inconsistent with position of the body.
B. Rigor (stage/intensity, location on the body, broken, inconsistent with the scene).
C. Degree of decomposition (putrefaction, adipocere, mummification, skeletonization, as appropriate).
D. Insect and animal activity.
E. Scene temperature (document method used and time estimated).
F. Description of body temperature (e.g., warm, cold, frozen) or measurement of body temperature (document method used and time of measurement).

Summary:

Documentation of post mortem changes in every report is essential to determine an accurate cause and manner of death, provide information as to the time of death, corroborate witness statements, and indicate that the body may have been moved after death.

6. Participate in Scene Debriefing

Principle: The scene debriefing helps investigators from all participating agencies to establish post-scene responsibilities by sharing data regarding particular scene findings. The scene debriefing provides each agency the opportunity for input regarding special requests for assistance, additional
information, special examinations, and other requests requiring interagency communication, cooperation, and education.

Authorization: Medical Examiner/Coroner Official Office Policy Manual; State or Federal Statutory Authority.

Policy: The investigator shall participate in or initiate interagency scene debriefing to verify specific post-scene responsibilities.

Procedure: When participating in scene debriefing, the investigator should:

A. Determine post-scene responsibilities (identification, notification, press relations, and evidence transportation).
B. Determine/identify the need for a specialist (e.g., crime laboratory technicians, social services, entomologists, OSHA).
C. Communicate with the pathologist about responding to the scene or to the autopsy schedule (as needed).

D. Share investigative data (as required in furtherance of the investigation).
E. Communicate special requests to appropriate agencies, being mindful of the necessity for confidentiality.

Summary:

The scene debriefing is the best opportunity for investigative participants to communicate special requests and confirm all current and additional scene responsibilities. The debriefing allows participants the opportunity to establish clear lines of responsibility for a successful investigation.

7. Determine Notification Procedures (Next of Kin)

Principle: Every reasonable effort should be made to notify the next of kin as soon as possible. Notification of next of kin initiates closure for the family, disposition of remains, and facilitates the collection of additional information relative to the case.

Authorization: Medical Examiner/Coroner Official Office Policy Manual;
State or Federal Statutory Authority.

Policy: The investigator shall ensure that next of kin is notified of the death and that all failed and successful attempts at notification are documented.

Procedure: When determining notification procedures, the investigator should:

A. Identify next of kin (determine who will perform task).
B. Locate next of kin (determine who will perform task).
C. Notify next of kin (assign person(s) to perform task) and record time of notification, or, if delegated to another agency, gain confirmation when notification is made.
D. Notify concerned agencies of status of the notification.

Summary:

The investigator is responsible for ensuring that the next of kin is identified, located, and notified in a timely manner. The time and method of notification should be documented. Failure to locate next of kin and efforts to do so should be a matter of record. This ensures that every reasonable effort has been made to contact the family.

8. Ensure Security of Remains

Principle: Ensuring security of the body requires the investigator to supervise the labeling, packaging, and removal of the remains. An appropriate identification tag is placed on the body to preclude misidentification upon receipt at the examining agency. This function also includes safeguarding all potential physical evidence and/or property and clothing that remain on the body.

Authorization: Medical Examiner/Coroner Official Office Policy Manual;
State or Federal Statutory Authority.

Policy: The investigator shall supervise and ensure the proper identification, inventory, and security of evidence/property and its packaging and removal from the scene.

Procedure: Prior to leaving the scene, the investigator should:

A. Ensure that the body is protected from further trauma or contamination (if not, document) and unauthorized removal of therapeutic and resuscitative equipment.

B. Inventory and secure property, clothing, and personal effects that are on the body (remove in a controlled environment with witness present).
C. Identify property and clothing to be retained as evidence (in a controlled environment).
D. Recover blood and/or vitreous samples prior to release of remains.
E. Place identification on the body and body bag.
F. Ensure/supervise the placement of the body into the bag.
G. Ensure/supervise the removal of the body from the scene.
H. Secure transportation.

Summary:

Ensuring the security of the remains facilitates proper identification of the remains, maintains a proper chain of custody, and safeguards property and evidence.

Section E: Establishing and Recording Decedent Profile Information

1. Document the Discovery History

Principle: Establishing a decedent profile includes documenting a discovery history and circumstances surrounding the discovery. The basic profile will dictate subsequent levels of investigation, jurisdiction, and authority. The focus (breadth/depth) of further investigation is dependent
on this information.

Authorization: Medical Examiner/Coroner Official Office Policy Manual;
State or Federal Statutory Authority.

Policy: The investigator shall document the discovery history, available witnesses, and apparent circumstances leading to death.

Procedure: For an investigator to correctly document the discovery history, he/she should:

A. Establish and record person(s) who discovered the body and when.
B. Document the circumstances surrounding the discovery (who, what, where, when, how).

Summary:

The investigator must produce clear, concise, documented information concerning who discovered the body, what are the circumstances of discovery, where the discovery occurred, when the discovery was made, and how the discovery was made.

2. Determine Terminal Episode History

Principle: Pre-terminal circumstances play a significant role in determining cause and manner of death. Documentation of medical intervention and/or procurement of ante mortem specimens help to establish the decedent's condition prior to death.

Authorization: Medical Examiner/Coroner Official Office Policy Manual;
State or Federal Statutory Authority.

Policy: The investigator shall document known circumstances and medical intervention preceding death.

Procedure: In order for the investigator to determine terminal episode history, he/she should:

A. Document when, where, how, and by whom decedent was last known to be alive.
B. Document the incidents prior to the death.
C. Document complaints/symptoms prior to the death.
D. Document and review complete EMS records (including the initial electrocardiogram).
E. Obtain relevant medical records (copies).
F. Obtain relevant ante mortem specimens.

Summary:

Obtaining records of pre-terminal circumstances and medical history distinguishes medical treatment from trauma. This history and relevant ante mortem specimens assist the medical examiner/coroner in determining cause and manner of death.

3. Document Decedent Medical History

Principle: The majority of deaths referred to the medical examiner/coroner are natural deaths. Establishing the decedent's medical history helps to focus the investigation. Documenting the decedent's medical signs or symptoms prior to death determines the need for subsequent examinations.
The relationship between disease and injury may play a role in the cause, manner, and circumstances of death.

Authorization: Medical Examiner/Coroner Official Office Policy Manual;
State or Federal Statutory Authority.

Policy: The investigator shall obtain the decedent's past medical history.

Procedure: Through interviews and review of the written records, the investigator should:

A. Document medical history, including medications taken, alcohol and drug use, and family medical history from family members and witnesses.
B. Document information from treating physicians and/or hospitals to confirm history and treatment.
C. Document physical characteristics and traits (e.g., left-/right-handedness, missing appendages, tattoos, etc.).

Summary:

Obtaining a thorough medical history focuses the investigation, aids in disposition of the case, and helps determine the need for a post mortem examination or other laboratory tests or studies.

4. Document Decedent Mental Health History

Principle: The decedent's mental health history can provide insight into the behavior/state of mind of the individual. That insight may produce clues that will aid in establishing the cause, manner, and circumstances of the death.

Authorization: Medical Examiner/Coroner Official Office Policy Manual;
State or Federal Statutory Authority.

Policy: The investigator shall obtain information from sources familiar with the decedent pertaining to the decedent's mental health history.

Procedure: The investigator should:

A. Document the decedent's mental health history, including hospitalizations and medications.
B. Document the history of suicidal ideations, gestures, and/or attempts.
C. Document mental health professionals (e.g., psychiatrists, psychologists, counselors, etc.) who treated the decedent.
D. Document family mental health history.

Summary:

Knowledge of the mental health history allows the investigator to evaluate properly the decedent's state of mind and contributes to the determination of cause, manner, and circumstances of death.

5. Document Social History

Principle: Social history includes marital, family, sexual, educational, employment, and financial information. Daily routines, habits and activities, and friends and associates of the decedent help in developing the decedent's profile. This information will aid in establishing the cause, manner, and circumstances of death.

Authorization: Medical Examiner/Coroner Official Office Policy Manual;
State or Federal Statutory Authority.

Policy: The investigator shall obtain social history information from sources familiar with the decedent.

Procedure: When collecting relevant social history information, the investigator should:

A. Document marital/domestic history.
B. Document family history (similar deaths, significant dates).
C. Document sexual history.
D. Document employment history.
E. Document financial history.
F. Document daily routines, habits, and activities.
G. Document relationships, friends, and associates.
H. Document religious, ethnic, or other pertinent information (e.g., religious objection to autopsy).
I. Document educational background.
J. Document criminal history.

Summary:

Information from sources familiar with the decedent pertaining to the decedent's social history assists in determining cause, manner, and circumstances of death.

Section F: Completing the Scene Investigation

1. Maintain Jurisdiction over the Body

Principle: Maintaining jurisdiction over the body allows the investigator to protect the chain of custody as the body is transported from the scene for autopsy, specimen collection, or storage.

Authorization: Medical Examiner/Coroner Official Office Policy Manual;
State or Federal Statutory Authority.

Policy: The investigator shall maintain jurisdiction of the body by arranging for the body to be transported for autopsy, specimen collection, or storage by secure conveyance.

Procedure: When maintaining jurisdiction over the body, the investigator should:

A. Arrange for, and document, secure transportation of the body to a medical or autopsy facility for further examination or storage.

B. Coordinate and document procedures to be performed when the body is received at the facility.

Summary:

By providing documented secure transportation of the body from the scene to an authorized receiving facility, the investigator maintains jurisdiction and protects chain of custody of the body.

2. Release Jurisdiction of the Body

Principle: Prior to releasing jurisdiction of the body to an authorized receiving agent or funeral director, it is necessary to determine the person responsible for certification of the death. Information to complete the death certificate includes demographic information and the date, time, and location of death.

Authorization: Medical Examiner/Coroner Official Office Policy Manual; State or Federal Statutory Authority.

Policy: The investigator shall obtain sufficient data to enable completion of the death certificate and release of jurisdiction over the body.

Procedure: When releasing jurisdiction over the body, the investigator should:

A. Determine who will sign the death certificate (name, agency, etc.).
B. Confirm the date, time, and location of death.
C. Collect, when appropriate, blood, vitreous fluid, and other evidence prior to release of the body from the scene.
D. Document and arrange with the authorized receiving agent to reconcile all death certificate information.
E. Release the body to a funeral director or other authorized receiving agent.

Summary:

The investigator releases jurisdiction only after determining who will sign the death certificate; documenting the date, time, and location of death; collecting appropriate specimens; and releasing the body to the funeral director or other authorized receiving agent.

3. Perform Exit Procedures

Principle: Bringing closure to the scene investigation ensures that important evidence has been collected and the scene has been processed. In addition, a systematic review of the scene ensures that artifacts or equipment are not inadvertently left behind (e.g., used disposable gloves, paramedical debris, film wrappers, etc.), and any dangerous materials or conditions have been reported.

Authorization: Medical Examiner/Coroner Official Office Policy Manual; State or Federal Statutory Authority.

Policy: At the conclusion of the scene investigation, the investigator shall conduct a post-investigative "walk through" and ensure the scene investigation is complete.

Procedure: When performing exit procedures, the investigator should:

A. Identify, inventory, and remove all evidence collected at the scene.
B. Remove all personal equipment and materials from the scene.
C. Report and document any dangerous materials or conditions.

Summary:

Conducting a scene "walk through" upon exit ensures that all evidence has been collected, that materials are not inadvertently left behind, and that any dangerous materials or conditions have been reported to the proper entities.

4. Assist the Family

Principle: The investigator provides the family with a timetable so they can arrange for final disposition and provides information on available community and professional resources that may assist the family.

Authorization: Medical Examiner/Coroner Official Office Policy Manual; State or Federal Statutory Authority.

Policy: The investigator shall offer the decedent's family information regarding available community and professional resources.

Procedure: When the investigator is assisting the family, it is important to:

A. Inform the family if an autopsy is required.
B. Inform the family of available support services (e.g., victim assistance, police, social services, etc.).
C. Inform the family of appropriate agencies to contact with questions (medical examiner/coroner offices, law enforcement, SIDS support group, etc.).
D. Ensure family is not left alone with body (if circumstances warrant).
E. Inform the family of approximate body release timetable.
F. Inform the family of information release timetable (toxicology, autopsy results, etc., as required).
G. Inform the family of available reports, including cost, if any.

Summary:

The interaction with the family allows the investigator to assist and direct them to appropriate resources. It is essential that families be given a timetable of events so that they can make necessary arrangements. In addition, the investigator needs to make them aware of what and when information will be available. Knowing the process and how the forensic meteorologist fits in can help in the investigation process. Evidence gathering should include environmental elements (pictures, sensor readings…)

Questions for Discussion

1. What is missing in Section 'A' of Chapter 7 – Investigative Tool and Equipment? See if you see anything that relates to your profession as a forensic meteorologist.

2. Invite the local Medical Examiner to class and discuss the varying cases they were involved in where working with a forensic meteorologist could have assisted in their investigation.

3. Participate in an autopsy – through the Medical Examiner, and learn what parts of the body are more susceptible to temperature variations, what organ holds the body temperature the longest. Although not a requirement for this text, it certainly makes our profession as a forensic meteorologist more interesting.

4. In reviewing Chapter 7, explain your theory as to why there's so little reference to 'weather' and 'weather elements' associated with a crime scene.

5. Some students may be EMT certified; find out if you are qualified (usually the geographically challenged will assign this to certain EMT's) to take a body temperature and if so, how many sample readings would you need to determine a rate of cooling?

Chapter 8

Aviation Accident Investigations

This chapter will give a brief overview as 99.9% of investigations will come from the NTSB. Having a familiarity with the concept and techniques will help you determine if this is a good career move for you.

As of this printing - The Forensic Services Program is the focal point of National Weather Service (NWS) support for transportation accident investigations and potential subsequent litigation. Much of the work is related to aviation accidents, but a considerable amount of effort is related to marine and other surface accidents. The program activity includes:

Ensuring that high-level National Weather Service and National Oceanic and Atmospheric Administration (NOAA) management are provided timely notification, and kept informed of transportation accidents of unusual interest.

The National Transportation Safety Board (NTSB) investigates aviation transportation accidents. Representing the National Weather Service at NTSB Public Hearings and preparing NWS personnel designated as witnesses to testify at Public Hearings. Responding to NTSB Safety Recommendations addressed to the NWS evolving from transportation accident investigations.

Maintaining an awareness of all investigative and/or legal activity evolving from weather-related transportation and pipeline accidents. Providing technical advice to Government attorneys defending suits against the Government evolving from transportation accidents.

Assisting Government attorneys in preparing NWS employees for testimony at legal proceedings. Responding to legal production, Freedom of Information Act (FOIA), and mission information requests for weather records documentation.

Ensuring the NWS records retention program meets agency litigation-support requirements.

NTSB Field Investigation

The NTSB is responsible for the investigation and determination of probable cause of all civil aircraft accidents, certain highway and railroad accidents, pipeline accidents in which there is a fatality or substantial property damage, and certain major marine casualties. Many of these transportation accidents are relatively low key, with the investigations handled locally by NTSB regional staff. Some are delegated to the Federal Aviation Administration (FAA).

However, after any fatal accident involving an air carrier, NTSB's national headquarters will conduct the field investigation, to determine the facts, conditions, and circumstances related to the event. A "go-team," headed by the NTSB Investigator-in Charge is at the accident scene within 24 hours of the accident. A field investigation will last anywhere from one to three weeks, depending on the scope of the accident in question. When the accident is suspected to be weather-related, the "go-team" will include an NTSB meteorologist and may include a representative from the NWS - normally a member of the Forensic Services staff. As a member of the Weather Group, the NWS representative assists in gathering appropriate "work" documents and interviewing NWS personnel, and participates in the development of the written factual record of the on-scene phase of the accident. In cases where there is no NWS representative on the "go-team," the NTSB meteorologist will contact a designated NWS regional official, usually the Regional Aviation Meteorologist (RAM), or equivalent, for assistance. The RAM will act as a liaison between the NTSB and NWS offices and personnel. A list of NWS liaisons is maintained and routinely updated by the Forensic Services program and passed on to the NTSB.

NTSB Public Hearing

The NTSB, at its discretion, may convene a public hearing as a part of its investigation into a transportation accident for the purpose of creating a public record of the facts, conditions, and circumstances relating to the accident. A public hearing usually occurs within two to three months after an accident. It normally lasts for a week or less (but may be considerably longer - the hearing evolving from a Trans World Airlines crash near Berryville, Virginia, in 1974, lasted for four weeks)!

The key players at NTSB Hearings are the NTSB Technical Panel, the Spokespersons for the various designated parties, and the witnesses. The NWS is normally a designated party in weather-related cases where the testimony of NWS employees is required. Commonly, NWS witness roster will include the duty surface weather and radar observers, the duty NWS office forecaster, and the duty Center Weather Service Unit meteorologist. Most of the questions addressed to a participating witness will originate from a member of the Technical Panel. Where NWS witnesses are concerned, this member will be the NTSB meteorologist assigned to the case. However, all of the spokespersons of the various designated parties, some of which may be preparing for subsequent litigation, are permitted to ask questions.

NTSB Aircraft Accident Report

Upon completion of its investigative activities, the NTSB publishes the Aircraft Accident Report. With respect to carrier accidents, this will occur anywhere from one to two years after the accident. The Report contains detailed factual information as well as an extensive analysis of the factors and circumstances surrounding the accident question. It also itemizes findings of the NTSB and assigns probable and contributing causes. The latter is the only part of the Report which cannot be used in litigation. It is more than a coincidence that potential legal activity is normally deferred until after the NTSB has published its Report.

The Aircraft Accident Report will commonly include multiple Safety Recommendations relating to issues surrounding the accident in question. These will be addressed to the appropriate Government agencies, as well as civil parties and organizations. Based on its investigations and findings, the NTSB will identify deficiencies and oversights, and make recommendations for remedial or corrective action, so as to prevent a similar reoccurrence in the future. The Forensic Services program is the NWS focal point for responding to Safety Recommendations relating the aviation weather operations and services.

Litigation Process

Overview

Legal activity in cases involving the NWS, from the time that administrative claims are presented to the agency, normally lasts for up to five years - although cases have gone on as long as ten years. During this period, Forensic Services staff may be required to produce hundreds of operational and administrative documents related (or determined to be related) to the case, and spend many hours dedicated to providing technical advice to Government attorneys and assisting in preparing NWS employees for testimony. It is noteworthy that the 46 cases producing administrative claims against the NWS during the period from 1980 through 1998 had total monetary attachments of almost 1.6 billion dollars! The enormous amount of time and resource investment in defense against these inflated claims is made not only to conserve U.S. treasury funds, but of equal importance, to prevent a court ruling which may adversely affect future NWS operations and/or its employees to exercise their best professional judgment.

Documentation Case Files

The Forensic Services Program establishes and maintains about 250 - 400 each year.

Each case represents a request or often several requests for weather information.

Response to weather documentation requests includes the retrieval, certification, and production of historic aviation and non-aviation (e.g. marine, surface, hydrology) records

Requests for weather documentation may come from private citizens; weather consultants; attorneys; and other federal, state, or local government agencies

Requests for weather data may be simple and straight forward while others may take a considerable effort and cooperation from many sources to fulfill

- By most of the case files handled by the Forensic Services are related to aviation accidents; however, a large number are related to marine and other surface accidents

- The public accounts for most of the customer requests. Primary Government customers are the NTSB, the FAA and the DOJ. Other miscellaneous customers include the US Coast Guard, offices within the Department of Commerce, and state and local law enforcement agencies.

Past weather records can be obtained in 2 ways:

For all non-certified records or certified climatological records (including radar images, satellite photos, surface analysis, buoy or ship reports), please contact:

[National Climatological Data Center](#)
151 Patton Avenue
Asheville, NC 28801-5001
(828) 271-4800

For CERTIFIED FORECASTS (or any requests that include the need for certified forecasts), the requests MUST be in writing, either via mail or fax. Please contact:

NWS Headquarters
ATTN: W/OS52
1325 East-West Highway
Silver Spring, MD 20910
(Courtesy: http://www.nws.noaa.gov/om/forensic.shtml)

Microbursts are a phenomena that has been the cause for many aviation accidents. Fernando Caracena and Dr. Fujita set out to show a classical situation of a microburst.

Collaborating with Dr. Fujita

Fernando Caracena, NOAA, Forecast Systems Laboratory, Boulder, Colorado

In 1976 Don Beran (of NOAA's Wave Propagation Laboratory) suggested that I write an article about the crash of Continental flight 426 as a contribution to a special windshear issue of the Bulletin of the American Meteorological Society (BAMS), which he was editing—perhaps collaborating with Dr. Fujita who was also preparing a BAMS article on the crash of Eastern Flight 66.

At Fujita's invitation, I flew up for over a week's work with him at his laboratory at the University of Chicago. By then, Allegheny Flight 121 had crashed while attempting to land at Philadelphia International Airport on the afternoon of 23 June 1976 as it emerged from a "wall of water." For over a week, we worked side by side, analyzing the three accidents using our combined data sets.

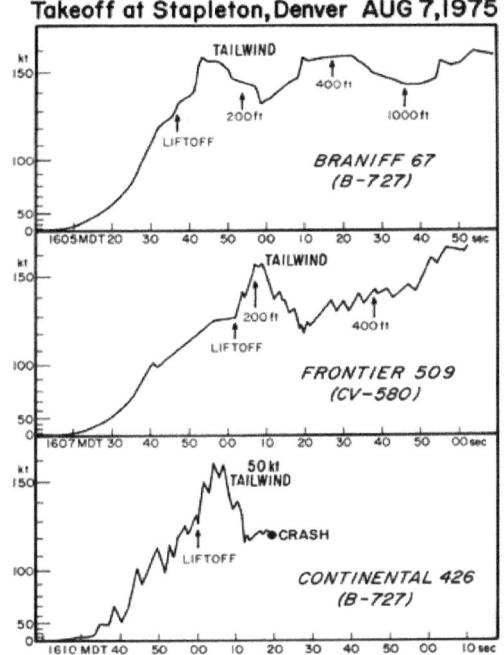

Fujita's weather analysis technique

Fujita made mesoscale weather analyses look like a leisurely drive down a country road on a fine autumn day. He told me, "The hardest part is collecting all the data." For Fujita, the rest was easy. ... In the end, Fujita wove a four-dimensional weather analysis, superposed on radar imagery that covered the accident time and location.

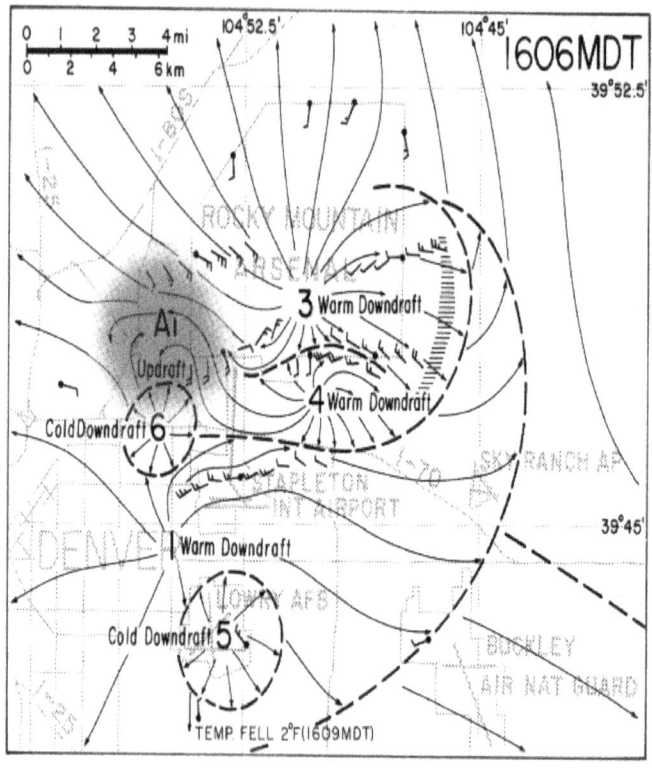

A small circular radar echo formed west of the airport 4 min before Continental 426 took off. Six minutes later, the echo had grown into a large spearhead that covered the north end of the airport. Meanwhile, the microbust was impacting on runway 35L.

Greg MacMaster   Environmental Forensics

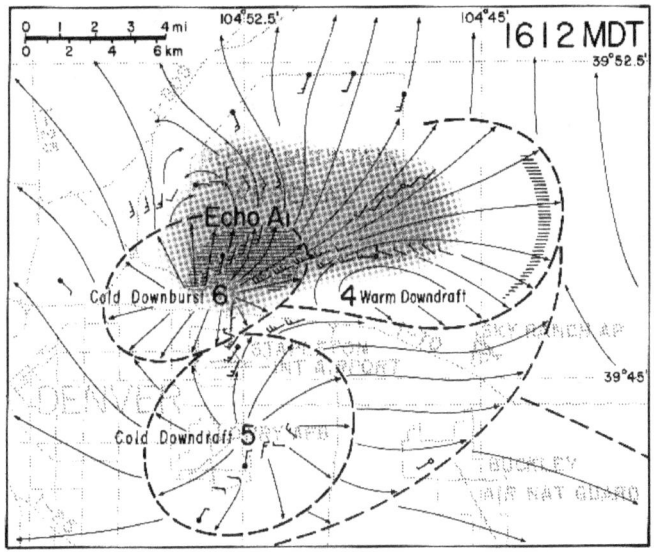

I was gratified to see that Fujita's analysis agreed with mine. But, his was much richer in detail. The valuable weather analysis techniques that I learned by collaborating with Dr. Fujita would serve me well in the future.

Historical importance of the three weather-related accidents

The years, 1975 and 1976, were landmarks in air safety history. First, Fujita identified the microburst as a threat to aviation. Second, our weather analysis of the Denver accident documented that a microburst was the cause of the crash. Finally, a third crash in a short span of time made clear the mounting threat of microbursts to aviation. The microburst was a beast that ate airplanes. Meanwhile, some aircraft weather experts testified that this beast could not exit. But, Fujita continued to compile a growing list of crashes involving microbursts that grew to 8 by 1985.

Wet and Dry Microbursts

(Fernando Caracena NOAA, Forecast Systems Laboratory, Boulder, Colorado)

Wet and dry microbursts-operational definitions

The field experiments of the 1980s and 90s (NIMROD, JAWS, MIST and others) revealed microburst wind gusts striking a network site with measurable rain, and others, without. Two of the three accidents of 1975 and 1976 involved heavy rain and another, only vanishing sprinkles.

NIMROD (Northern Illinois Meteorological Research on Downburst, Chicago, 1978-79), JAWS (Joint Airport Weather Studies, Denver, 1982-86)

MIST (Microburst and Severe Thunderstorm project, Huntsville, Ala., 1986-88).

Dry storms can exist in environments having high lapse rates. Braham (1985) and W. R. Krumm (1954) describe a dry, high-based thunderstorm and its environment common in the western United States, which produces lighting, almost no rain, but lots of strong, sometimes damaging wind.

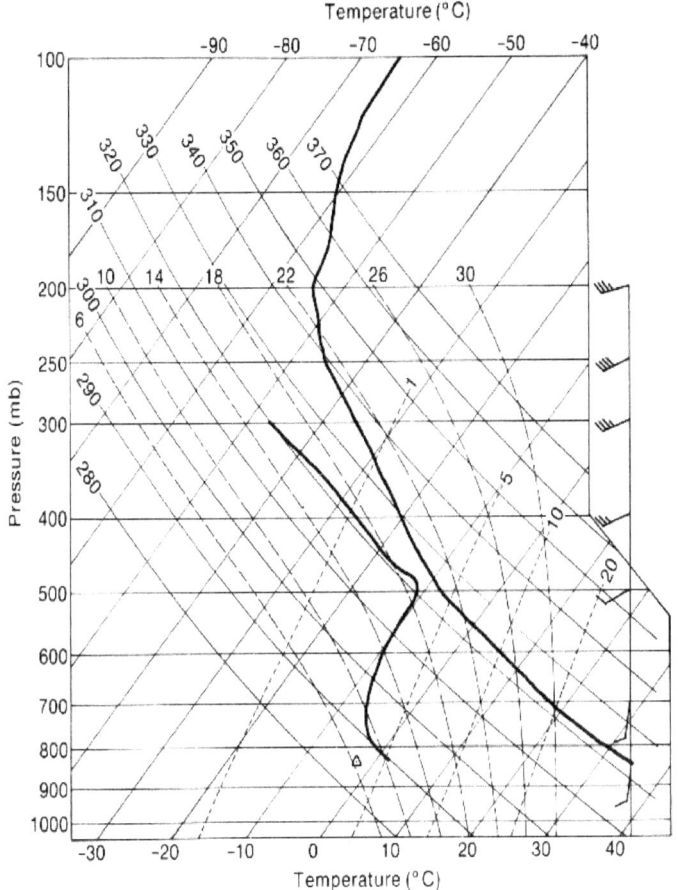

Brown et al. (1982) identify an even more extreme, dry storm environment (Fig. 2 a), in which the dry static stability almost vanishes and just enough moisture stratification remains to support cumulonimbi without lightning, but with just enough precipitation to drive severe downdrafts. Weak updrafts and cold condensation temperatures grow precipitation mostly as lightly rimed ice crystals, which falling through the melting level, become a spray of small raindrops that evaporate rapidly.

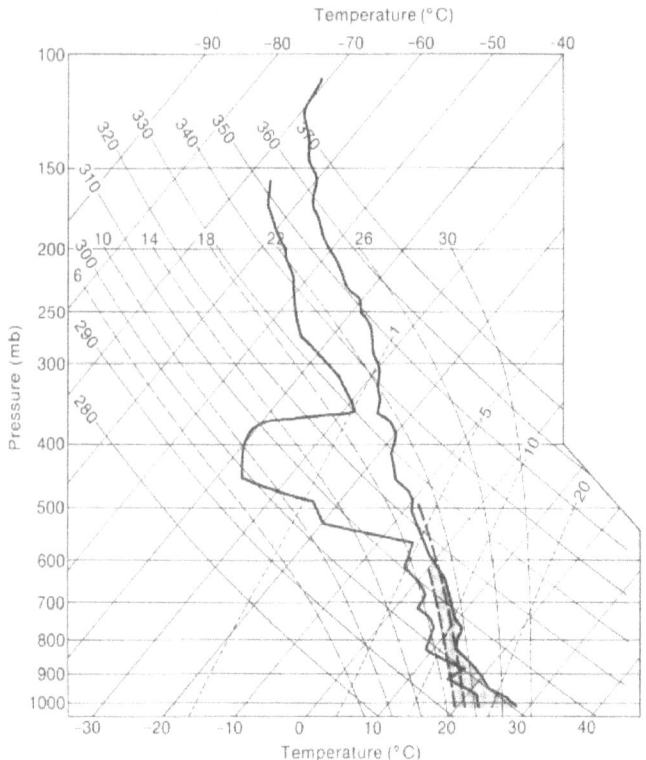

A local sounding taken at the Florida Area Cumulus Experiment three hours before a wet microburst struck this site reveals that the source of the strong downdraft was an elevated dry layer containing air potentially cold enough to inject a cool downdraft into the precipitation column at a high level from where it could accelerate to the surface.

Microburst information Courtesy: Fernando Caracena NOAA, Forecast Systems Laboratory, Boulder, Colorado .PhD, Physicist
Mesoscale Applications Group - Forecast Research Division - Forecast Systems Laborator
NOAA

Thunderstorms, Icing and Turbulence are a determent to aviation. If thunderstorms are forecast, severe icing and severe turbulence is implied and flight should be restricted into these storms. However, there are some cloud types that suggest severe up and downdrafts without showing the characteristics. I'm referring to the Cb cloud. A cumulonimbus cloud (without the anvil) has the same characteristics of a Cb with an anvil. Strong up and down drafts, severe turbulence and of course, severe icing are hazards pilots try to stay away from.

Forecasting Icing Using the -8D Rule

1. At significant levels on the Skew-T, determine temperature and dew-point spread. The result is the "D" value.

2. Multiply the "D" value by -8 and plot at the significant level.

3. Connect the plotted points by drawing a straight line from point to point.

4. Where line crosses right of temperature trace, clouds are supersaturated with respect to ice and icing is possible.

Icing Intensity (White Sands Missile Range)

$L = (1/2.87) * (w_o - w) * (P/t)$. W_o and W is the mixing ratio at cloud base and at level in question. P is the millibar pressure level in question and the T is the temperature in Kelvin (273ºC)

Other Rules for Forecasting Moderate or Greater Clear Air Turbulence

1. Whenever winds aloft, vertical shear, equals or exceeds 30 knots per 5000 feet.

2. For jet stream stronger than 110 knots at core, particularly in sloping tropopause above core.
3. If 300mb 20 knot isotachs are less than 60 nautical miles apart.
4. In entrance zone between two confluent jet cores in region where cores are 3-5º latitude or less apart.
5. In sharp, V-shaped troughs which slope rapidly with elevation below 300mb.
6. In "neck" of cut-off lows.
7. To right of strong anticyclonically curved jet streams.
8. Vertical extent is no higher than 2000 feet above tropopause and within layers described above.

9. Strong horizontal temperature gradients (10°C/100 miles) signify strong vertical wind shear and turbulence.

Turbulence Chart for Aviation (According to Aircraft Category)

Category I	Category II	Category III	Category IV
N	N	N	N
(L)	N	N	N
Light	(L)	N	N
L-(M)	L	(L)	N
M	L-(M)	L	(L)
M-(S)	M	L-(M)	L
S	M-(S)	M	L-(M)
S-(X)	S	M-(S)	M
X	S-(X)	S	M-(S)
X	X	S-(X)	S
X	X	X	S-(X)
X	X	X	X

L = Light   M = Moderate   S = Severe

This chart displays the severity of turbulence according to the type of aircraft that fits within that category. Category I aircraft depicts lighter aircraft (Cessna 152 for example) all the way to category IV which includes the heavier aircraft, jet fighters, larger transport carriers. *(provided by Greg MacMaster)*

SEVERE THUNDERSTORMS AND TORNADOES

Temperature The most vigorous storms occur in air having a morning subsidence inversion, a characteristic associated with wind-shear –both directional and speed.

Moisture The strongest storms require drier air above the ridge or a marked dry source to the windward side in a position to intrude into or over the moist ridge. This dry slot is best found at 700mbs.

Winds Strong mid-level winds are required with sharp horizontal wind-shears. Moderate to strong the low level flow is required for major storm family type outbreaks. The intersection of the maximum low-level winds with a warm front or old squall line is a frequent development area. It is extremely desirable for the mid-level and low-level jets to intersect, since the tornado activity major axis is frequently determined by the location and movement of this intersection.

Vorticity Strong PVA (Positive Vorticity Advection) interacting with a dryline leads to a much higher probability of igniting intense to severe thunderstorms. Keep in mind that PVA is a relative number and the strength associated with a number will change the next model run.

Stability of the air column depends largely on the low-level moisture and temperature advection.

Upper-level Jet Cyclogenesis can result when there is a dynamic coupling of the low-level jet's right-front with the upper-level jet's left-front quadrant. There is a significant correlation between severe storm development when a southerly low-level jet is bisected at an angle of 30º or more.

Low-level Jet The highest upward lifting rate (enhancing low-level warm advection) is associated with the nose and the right-front quadrant of the low-level jet.

Low-level Moisture is best determined from the sounding's analysis of the lowest 3,000 feet using the moist layer's mean dew point. Or, take the 850mb and surface dew point temperature averages.

850mb Temperature Pattern The maximum temperature ridge is located to the west or southwest (upwind) of the low-level moisture ridge. It is extremely important to locate the highest rate of warm advection area. The low-level thermal ridge is associated the with highest lapse rates but is low in moisture, while the low-level moisture axis downstream has marginal conditional instability in levels above it. When these regions overlap, the potential for intense convection is greatest. The vertical totals index helps to diagnose steep lapse rate regions.

700mb No-change Line connects points of no advective temperature changes. This line separates warm air advection areas from cold air advection areas. It is coincident with the 850mb warm ridge. It has been observed that the advancement of the no-change line's leading edge into a position ahead of a pronounced mid-level trough is usually associated with deepening or intensification of surface low centers.

700mb Dry Air Intrusion Consider 50 percent relative humidity, dew point less than 0°C, or a temperature/dew point spread of more than 6°C at 700mb as dry. The most favorable relationship between the low-level and 700mb flow occurs when the winds veer with height.

Surface Pressure below 1013mb, and never above 1018mb.

Surface Dew Point The incidence of tornadoes is infrequent at dew-point temperatures below 13°C.

Twelve-hour Surface Pressure Falls reliably indicate major changes or trends in the surface pressure pattern and reflect important changes aloft. In the most productive situations, it appears that widespread pressure falls are less desirable than a pattern of more concentrated falls.

Winds - Veering and Backing

The wind veers when it changes direction clockwise. Example: The surface wind is blowing from 270°. At 2000 feet it is blowing from 280°. It has changed in a right-hand, or clockwise, direction.

The wind backs when it changes direction anti clockwise. Example: The wind direction at 2000 feet is 090°. at 3000 feet 085°. It is changing in a left-hand, or anti-clockwise, direction.

In a descent from several thousand feet above the ground to ground level, the wind will usually be found to back and also decrease in velocity, as the effect of surface friction becomes apparent. Likewise, in a climb from the surface to several thousand feet AGL, the wind will veer and increase.

At night, surface friction decreases as surface cooling reduces the eddy motion of the air. Surface winds will back and decrease. During the day, as surface friction intensifies, the surface winds will veer and increase.

Winds – Wind Shear

Wind shear is the sudden tearing or shearing effect encountered along the edge of a zone in which there is a violent change in wind speed or direction. It can exist in a horizontal or vertical direction and produces churning motions and consequently turbulence. Under some conditions, wind direction changes of as much as 180 degrees and speed changes of as much as 80 knots have been measured.

The effect on airplane performance of encountering wind shear derives from the fact that the wind can change much faster than the airplane mass can be accelerated or decelerated. Severe wind shears can impose penalties on an airplane's performance that are beyond its capabilities to compensate, especially during the critical landing and take-off phase of flight.

Winds - At Cruising Flight

In cruising flight, wind shear will likely be encountered in the transition zone between the pressure gradient wind and the distorted local winds at the lower levels. It will also be encountered when climbing or descending through a temperature inversion and when passing through a frontal surface. Wind shear is also associated with the jet stream (see below). Airplanes encountering wind shear may experience a succession of updrafts and downdrafts, reductions or gains in headwind, or wind shifts that disrupt the established flight path. It is not usually a major problem because altitude and airspeed margins will be adequate to counteract the shear's adverse effects. On occasion, however, the wind shear may be severe enough to cause an abrupt increase in load factor, which might stall the airplane or inflict structural damage.

Winds - Near the Ground

Wind shear, encountered near the ground, is more serious and potentially very dangerous. There are four common sources of low level wind shear: thunderstorms, frontal activity, temperature inversions and strong surface winds passing around natural or manmade obstacles.

Frontal Wind Shear. Wind shear is usually a problem only in fronts with steep wind gradients. If the temperature difference across the front at the surface is 5°C or more and if the front is moving at a speed of about 30 knots or more, wind shear is likely to be present. Frontal wind shear is a phenomenon associated with fast moving cold fronts but can be present in warm fronts as well.

As a pilot, flying in weather that is pushing the limits of my ability creates stress and alters my state of mind. Needless to say, causes of accidents can point to one variable or could be innumerable. To be a member of the NTSB and investigate aviation accidents, you must be a licensed pilot.

Questions for Discussion

1. What qualifications are needed to work for the NTSB on aircraft accidents?
2. Who has the authority to make decisions in flight? Air Traffic Controller, Approach and Control? Ground Control? FAA? NTSB? Pilot? Discuss with class.
3. What are the main weather factors that affect flight?
4. If a pilot is vectored in for a lengthy approach on final and notices that he'll fly right into the base of a thunderstorm, what choices does the pilot have?
5. A good class project would be to visit a flight simulator and have the students fly (with an instructor) and see how severe weather elements affect flight.
6. If you have access to specific weather data that is not part of an officially approved site (such as ASOS), can you provide this to the NTSB in hopes to aid their investigation? Discuss.

Chapter 9
Construction

There are many facets to construction where the atmosphere plays a role. If proper planning of anticipated atmospheric conditions can be performed, risk and liability are reduced. But, there are many occasions where we don't have the for-sight to accommodate such factors since they have very little knowledge in climatology.

A case where there would be a factor; increased risk in roof damage would be in the shingle placement along the ridge of a roof. If the prevailing winds are from the west, then placement of the cut shingles should be placed from an easterly position and work your way to the west. The layering of the shingles would favor the wind/rain from the west and there would be less water damage or worn shingles in the future.

Digging a basement in a wet area: Consider the water table prior to building a basement. If it's dry, you can't assume that you'll have an easy dig as water could "hydraulic", or work its way up through the floor. If a study was performed to determine average rainfall and drainage as well as type of soil, clay as opposed to a sandy sediment, this could have an effect to the overall performance of your dwelling and future problems.

Water damage in a real estate transaction: Was it a one time, heavy rain event, or was it over a period of years (seasonal)? Did the previous owner know about the annual damage and hide it from the new owner? As a meteorologist, out job is to find – through whatever means possible – what may have led to the damage. Although most meteorologists may have very little background in construction, there are ways to find the cause even if there was no significant rain to lean on.
Take the placement of gutters for example. If a downspout didn't direct the water away from the structure, the excess water off the roof would run a path of least resistance, along the basement to a crack and wind up in the interior room.

There isn't enough room in the book to list all the problems that could develop. But we'll attempt to define a few.

Wind Issues

The forces imparted by wind onto buildings and other structures govern their design in many countries. Buildings are designed for strength, safety, and the comfort of the occupants in resisting both extreme wind storms (such hurricanes/typhoons) and commonly occurring winds (such as gradient winds, downbursts, monsoonal storms and Chinook winds). On the other hand, environmental issues, such as the transportation of airborne pollutants, may come to the fore during times of calm or low winds.

Wind is the movement of air relative to the surface of the earth. It's caused by solar heating of the surface of the Earth, which in turn causes pressure differences in the Earth's atmosphere, and forces generated by the rotation of the Earth.

Some common types of wind events:

Monsoons – are thermally produced seasonal winds that develop most strongly in and around the vast land mass of Asia.

Tropical Cyclones (also called Hurricanes and Typhoons) – are intense cyclonic storms driven by the latent heat of oceans, generally developing at latitudes between 5 and 20 degrees.

Thunderstorms – caused by the upwards movement of warm, moist air which then cools rapidly and starts to sink, condensation then produces heavy precipitation (rain or hail) which falls, dragging cold air downwards forming a downburst of wind.

Tornadoes – are a funnel-shaped vortex of air producing the most powerful of all winds on Earth.

Wind engineering uses a multifaceted approach to engineer a solution to wind issues and problems, and may involve various aspects of:

- The probability and statistics of the wind,
- Structure, and the environment;
- Meteorology,
- Fluid mechanics
- Structural dynamics.

The boundary layer wind tunnel has become the primary tool for use in studying wind issues and problems related to civil engineering applications. Wind tunnels are used to simulate the lowest layer of the atmosphere, known as the atmospheric boundary layer.

Delay Claims in Construction Cases: Common Pitfalls

Some of the most common disputes in construction cases relate to delay. As Scott Aftuck explains below, there are many things to consider. Delay claims tend to be some of the least understood and frequently confusing claims in the construction field. A clear understanding of the basic elements necessary to prove delay claims is invaluable in the processing of complex construction claims.

Much as it sounds, a delay claim on a construction project relates to a period of time for which the project has been extended or work has not been performed due to circumstances which were not anticipated when the parties entered into the construction contract. The most common causes of delay on a project include: differing site conditions; changes in requirements or design; weather; unavailability of labor, material or equipment; defective plans and specifications; and interference by the owner. Such delays will often force a contractor to extend its schedule to complete the work required under the contract, as well as to incur additional costs in the performance of said work. Generally, these costs may include: the costs of maintaining an idle workforce and equipment; unabsorbed office overhead; lost efficiencies; and general conditions. However, in order to receive an extension of time for project completion, or to recover additional costs, the contractor must meet a number of prerequisites.

A delay must be excusable in order to be the basis for an extension of time or additional compensation. Categories of excusable delay are often determined in the contract and typically involve matters beyond the control of the contractor. Examples of excusable delay include design errors and omissions, owner initiated changes, unanticipated weather, and acts of GOD. A non-excusable delay is a delay for which the contractor has assumed the risk under the contract. Oftentimes, even if a delay appears to be excusable, it will be the responsibility of the contractor if it was foreseeable; it could have been prevented but for the acts of the contractor, or it was caused by the negligence of the contractor.

Delays may be further classified into compensable and non-compensable delays. If a delay is compensable, the contractor is entitled to recover compensation for the costs of the delay in addition to time extensions to complete the project. Most contracts will include classes of delay which are compensable. The general rule, however, is that if the delay could have been avoided by due care of one of the parties, the party which did not exercise such care is responsible for the additional costs.

The contractor may also be liable for the negligent acts of its subcontractors. If the negligent subcontractor is in the chain of privity with the contractor, the contractor cannot recover delay damages from the owner as those delays are the responsibility of the contractor. However, if the subcontractor has a direct contractual relationship with the owner of the project, the contractor most likely will be able to recover damages as it was not in a position to prevent the delay. Additionally, in order to recover damages, a contractor must show a link between the delay and the resultant damage. Simply stating that there was a delay is not sufficient without showing a nexus between the delay and the damages.

Even if it is able to meet the foregoing criteria, a contractor will not be entitled to recovery if there is a concurrent delay effecting completion of the project. A concurrent delay may be defined as a second, independent delay occurring during the same time period as the delay for which recovery is sought. If the party seeking increased compensation is ultimately responsible for the concurrent delay, he may not be able to recover any compensation for the initial delay. Some courts, will allow the aggrieved party to attempt to apportion the responsibility for delay, thusly allowing compensation to the contractor for the period of delay which was not its responsibility. See, e.g., Raymond Constructors v. United States, 411 F.2d 1227 (Ct. Cl. 1969). However, apportionment of delay is often difficult due to inadequate project documentation of the various delays. The best time in which a contractor can apportion delay is while the project is ongoing. Courts tend to find analyses made concurrent with the delays to be more reliable than after the fact analyses.

Greg MacMaster   Environmental Forensics

A contractor can do a number of things to make it easier for it to recover delay damages incurred on a project. The contractor should make sure the construction contract clearly defines items which the contractor will be able to recover. Additionally, each and every delay incurred should be well documented during the course of the project. Notice that the delay is impacting the contractor should be given to the party with which it is in privity. Finally, if there is any portion of delay for which the contractor is responsible, it should seek to apportion the overall delay between the items it is responsible for and those for which it has no responsibility.
Courtesy of (Scott A. Aftuck of Haese, LLC)

Has El Nino Adversely Impacted Your Construction Project?

This year's winter rains have impacted almost every contractor. The abnormal weather conditions have delayed most projects and adversely impacted many contractors' productivity. Many contractors now ponder what rights they have to recover the additional costs that they have incurred as a result of the inclement weather. This article shall review what facts a contractor must establish to recover compensation for adverse weather conditions.

Weather Related Delay Claims-

In order to prevail on a delay claim, the party asserting the claim must prove that the delay was excusable, compensable and critical. If the contractor fails in establishing all three elements, it will not be able to recover additional compensation. However, if the contractor establishes that the delay was excusable and critical, it will be entitled to an extension of time.

Generally, most contracts do not excuse all weather related delays. Typically, the contract will only excuse rain delay days that were not foreseeable. Many government contracts now include an addendum that sets out the average rainfall for the anticipated construction period. For example, if during the scheduled period of construction, there were more than twenty days in which there was measurable rainfall, the contractor is only excused from performance on days with measurable rain after the twentieth day. In essence, these types of contract clauses require the contractor to assume the risk for all but unforeseeable weather-caused delays. Thus, it is necessary to first review the construction contract to determine how weather related delays are treated.

In most contracts, adverse weather delays are generally not compensable, but they will entitle a contractor to an extension of time. Nonetheless, if another preceding compensable delay extended the construction time into an adverse weather period, the weather related delays will be compensable. Similarly, if the owner requires the contractor to accelerate his performance to avoid adverse weather or to recapture the weather related delays then the contractor would be entitled to compensation. An example of additional compensation to recapture time lost to a rain delay is presently occurring at the San Francisco Airport. The Airport Commissioners are currently considering offering over nine million dollars in incentive, if the contractors can overcome the rain delays and complete the projects on time.

The final element of a weather related delay claim is establishing that it was a critical delay. Simply stated, the contractor must establish that the weather related delay adversely impacted the completion date for the project. Clearly, the completion date will not be impacted by abnormal rainfall when the only work that is occurring during the period is interior work. On the other hand, framing work would be impacted by abnormal rainfall. Thus, an analysis of how the work is impacted by the adverse weather is necessary.

Weather Related Disruption Claims--

While inclement weather may not delay the project, it could adversely impact the contractor's performance. For example, the adverse weather could result in an uneven labor force, inefficient labor force, out-of-sequence work, trade stacking, out-of-sequence material deliveries, and/or a loss of project momentum.

The determination of the right to compensation for a weather disruption claim is generally the same as that made for a weather related delay claim.

To prove a disruption claim, it will be necessary to identify the impacted activities. Specifically, the contractor should make contemporaneous records which establish which activities were impacted and the additional labor and costs associated with the disruption. If the contractor has failed to keep such contemporaneous records, there are other methods for proving the claim which include an engineered analysis, modified total cost, and/or a comparison analysis.

The disruption claim typically includes the additional direct and indirect costs for the impacted activity. The direct costs would include the increased labor, material and equipment costs. The indirect costs can include home office overhead and site overhead.

Documenting the Claim---

Obviously, any determination of an excusable weather delay or disruption will require proof that the work was impacted by the weather. Climatological data can be obtained from reports prepared by the National Atmospheric and Oceanic Administration. The reports can be found in most public libraries or on the Internet. In addition to establishing the actual number of rain days and the amount of rainfall, it is also necessary to prove that the contractor's work was in fact impacted by the delay.

The contractor should also keep daily job logs that document the weather conditions. The daily log should identify what trades, if any, were impacted by the adverse weather. The log should also set forth how the weather disrupted the project's performance.

In closing, if you properly documented the adverse weather this winter, *El Nino* storms may not have been as disastrous as originally believed. - *Courtesy William C. Last, Jr.*

Questions for Discussion

1. What lengths would you go to acquire weather data in a 'weather data sparse area' to obtain as much information as possible? Discuss all the possibilities.

2. Explain how you would verify or initialize non-official data to 'official data'. Would this help in your analysis? Why?

3. What product(s) would you review for a request of heavy rainfall that is over 20 miles away from a certified weather station?

4. Discuss with a contractor the possible loss of money (from workforce and material) when weather does not cooperate with their project.

5. If a contractor wants to pour 60 yards of cement and the forecast is expected to be 34 degrees and .75" of rain, - find out if he can continue his work OR if he did his work, what the potential loss would be in dollars (by not knowing what the forecast was expected to be).

6. If a flat rook has 36 inches of fresh snow (snow ratio 1/15) and in 2 weeks there's only 4 inches of snow (snow ratio of 1/3) on the roof after a mild spell, what would the implications be if that snow is not removed?

Chapter 10

Dendrochronology – Tree Dating

You receive a call and are asked if there was any flood data for a particular area that may have been recorded. You research data from your known sources and you come up with a moderate rainstorm, but nothing else. Do you have sufficient data, enough to provide the requestor? Can you help in his attempt to find additional information? Do you know who could help answer their questions? Sometimes it's not what you know – but who you know that can make or break your business as a forensic meteorologist. At the very least you should try to collaborate with other forensic professionals to find an answer.

Dendrochronology is a method of scientific dating based on the analysis of tree rings. Because the width of annular rings varies with **climatic conditions**, laboratory analysis of timber core samples allows scientists to reconstruct the conditions that existed when a trees' rings developed. By taking thousands of samples from different sites and different strata within a particular region, researchers can build a comprehensive historical sequence that becomes a part of the scientific record. Such master chronologies are used by archaeologists, climatologists, and others.

The science that uses annual tree rings dated to their exact year of formation for dating historical and environmental (weather) events. Trees are sensitive to both natural (precipitation and temperature patterns) events that trigger certain responses in the vigor of the tree as seen in its growth rate. In most geographic regions, climate patterns in any year cause a response by trees in the volume of wood the tree produces, and often leave indelible "fingerprints" in certain physical and chemical properties of the wood. These fingerprints can be seen in the varying widths of tree rings. In some years, environmental conditions may be favorable for tree growth, allowing trees to produce greater volumes of wood. In other years, climate conditions may be generally unfavorable for tree growth, causing a reduction in the volume of wood produced.

Dendrochronology has become a useful tool in many areas of research. Dendroarcheology uses tree rings to date wood material from archeological sites or artifacts. In instances where submerged logs (in the cold fresh waters of the Great Lakes) have been excavated for use in furniture, research is conducted to determine the dating of the log. The older the wood, the more expensive the furniture (or other product) is on the open retail market. In dendroclimatology climatic information is mathematically extracted from the tree-ring record and reconstructed back in time for the length of the tree-ring record. That data is also compared to data currently on file for accuracy and to fill in where some data is missing. Dendroclimatologists also use tree-ring records to quantify the rising levels of atmospheric carbon dioxide to better understand climate variations.

Dendrohydrology uses tree-ring data to investigate and reconstruct hydrologic properties, such as streamflow and riverflow, runoff, and past lake levels.

Dendrochronology has three main areas of application: paleoecology, where it is used to determine certain aspects of past ecologies (most prominently climate); archaeology, where it is used to date old buildings, etc.; and radiocarbon dating, where it is used to calibrate radiocarbon ages.

In some areas of the world, it is possible to date wood back a few thousand years, or even many thousands. In most areas, however, wood can only be dated back several hundred years, if at all.

History

Dendrochronology was developed during the first half of the 20th century originally by the astronomer A. E. Douglass, the founder of the Laboratory of Tree-Ring Research at the University of Arizona. Douglass sought to better understand cycles of sunspot activity and reasoned (correctly) that changes in solar activity would affect climate patterns on earth which would subsequently be recorded by tree-ring growth patterns (i.e., sunspots → climate → tree rings).

Growth rings

Growth rings also referred to as tree rings or annual rings, can be seen in a horizontal cross section cut through the trunk of a tree. Growth rings are the result of new growth in the vascular cambium, a lateral meristem, and are synonymous with secondary growth. Visible rings result from the change in growth speed through the seasons of the year, thus one ring usually marks the passage of one year in the life of the tree. The rings are more visible in temperate zones, where the seasons differ more markedly.

The inner portion of a growth ring is formed early in the growing season, when growth is comparatively rapid and is known as "early wood" or "spring wood" or "late-spring wood". The outer portion is the "late wood" (and has sometimes been termed "summer wood", often being produced in the summer, though sometimes in the autumn) and is denser. "Early wood" is used in preference to "spring wood", as the latter term may not correspond to that time of year in climates where early wood is formed in the early summer (e.g. Canada) or in autumn, as in some Mediterranean species

The growth rings of an unknown tree species, at Bristol Zoo, England.

Many trees in temperate zones make one growth ring each year, with the newest adjacent to the bark. For the entire period of a tree's life, a year-by-year record or ring pattern is formed that reflects the climatic conditions in which the tree grew. Adequate moisture and a long growing season result in a wide ring. A drought year may result in a very narrow one. Alternating poor and favorable conditions, such as mid summer droughts, can result in several rings forming in a given year. Missing rings are rare in oak and elm trees—the only recorded instance of a missing ring in oak trees occurred in the year 1816, also known as the 'Year Without a Summer'. Trees from the same region will tend to develop the same patterns of ring widths for a given period.

These patterns can be compared and matched ring for ring with trees growing in the same geographical zone and under similar climatic conditions. Following these tree-ring patterns from living trees back through time, chronologies can be built up, both for entire regions, and for sub-regions of the world. Thus wood from ancient structures can be matched to known chronologies (a technique called cross-dating) and the age of the wood determined precisely. Cross-dating was originally done by visual inspection, until computers were harnessed to do the statistical matching.

Pinus taeda Cross section showing annual rings, Cheraw, South Carolina.

To eliminate individual variations in tree ring growth, dendrochronologists take the smoothed average of the tree ring widths of multiple tree samples to build up a ring history. This process is termed 'replication'.

A tree ring history whose beginning and end dates are not known is called a 'floating chronology'. It can be anchored by cross-matching a section against another chronology (tree ring history) whose dates are known. Fully anchored chronologies which extend back more than 10,000 years exist for river oak trees from South Germany.

Another fully anchored chronology which extends back 8500 years exists for the bristlecone pine in the Southwest US (White Mountains of California). Furthermore, the mutual consistency of these two independent dendrochronological sequences has been confirmed by comparing their radiocarbon and endrochronological ages. In 2004 a new calibration curve INTCAL04 was internationally ratified for calibrated dates back to 26,000 Before Present (BP) based on an agreed worldwide data set of trees and marine sediments.

Sampling and dating

Pine stump showing growth rings.

Timber core samples measure the width of annual growth rings. By taking samples from different sites and different strata within a particular region, researchers can build a comprehensive historical sequence that becomes a part of the scientific record; for example, ancient timbers found in buildings can be dated to give an indication of when the source tree was alive and growing, setting an upper limit on the age of the wood. Some genera of trees are more suitable than others for this type of analysis. Likewise, in areas where trees grew in marginal conditions such as aridity or semi-aridity, the techniques of dendrochronology are more consistent than in humid areas. These tools have been important in archaeological dating of timbers of the cliff dwellings of Native Americans in the arid Southwest.

A benefit of dendrochronology is that it makes available specimens of once-living material accurately dated to a specific year to be used as a calibration and check of radiocarbon dating, through the estimation of a date range formed through the interception of radiocarbon (B.P., or 'B'efore 'P'resent, where present equals 1950-01-01) and calendar years. The bristlecone pine, being exceptionally long-lived and slow growing, has been used for this purpose, with still-living and dead specimens providing tree ring patterns going back thousands of years. In some regions dating sequences of more than 10,000 years are available.

There are many obstacles in dating tree rings, however, including some species of ant which inhabit trees and extend their galleries into the wood, thus destroying ring structure.

Similar seasonal patterns also occur in ice cores and in varves (layers of sediment deposition in a lake, river, or sea bed). The deposition pattern in the core will vary for a frozen-over lake versus an ice-free lake, and with the fineness of the sediment. These are used for dating in a manner similar to dendrochronology, and such techniques are used in combination with dendrochronology, to plug gaps and to extend the range of the seasonal data available to archaeologists and underwater or terrestrial archaeologist. Using this method can date an extreme event like a tsunami that claimed hundreds of thousands of lives hundreds of years ago. A sediment layer that was fairly thick could indicate a massive shift in the climate which stayed for a few hundred years before receding. This type of dating can either support or discount the Global warming debate which has captured so much attention recently.

While archaeologists can use the technique to date the piece of wood and when it was felled, it may be difficult to definitively determine the age of a building or structure that the wood is in. The wood could have been reused from an older structure, may have been felled and left for many years before use, or could have been used to replace a damaged piece of wood. You need to know when the tree was cut to make an accurate determination.

European chronologies derived from wooden structures found it difficult to bridge the gap in the 14th century when there was a building hiatus which coincided with the Black Death. Other plagues which were less well recorded also appear in the record.

In areas where the climate is reasonably predictable, trees develop annual rings of different properties depending on weather, rain, temperature, etc. in different years. These variations may be used to infer past climate variations.

Biosolids and Fertilizer

Greg MacMaster — Environmental Forensics

You've learned you can tell the age of a tree by looking at a cross section of the tree and counting the growth rings. The Douglas-fir tree on the next page was cut when it was about 30 years old. When it was 20 years old, the tree was fertilized with biosolids, resulting in the wider growth rings. Using biosolids can lead to impressive growth response in trees, especially in areas of the Pacific Northwest, where lack of nitrogen in native soils limits tree growth. Biosolids increases growth rates to match those in trees found on more fertile soils. Without knowing the effect of biosolids, one would think there was an impressive amount of rain over the course of 5+ years.

The other major limiting factor for tree growth is water. The organic matter in biosolids helps improve the ability of the soil to hold water and keep it available for the forest vegetation during the summer.

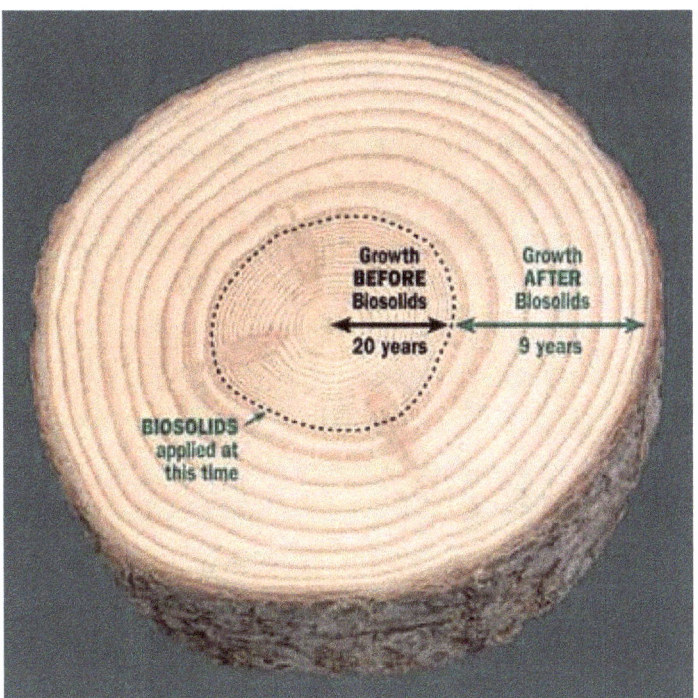

Greg MacMaster   Environmental Forensics

Courtesy: Biosolids Management Program

Summary:

While the majority of tree ring research is placed on the dendrochronologist, it's reassuring to know that there are other ways to find missing data instead of traditional sources like NCDC and local records. The true Forensic Meteorologist will investigate all options and bring in the professionals from other sciences to assist in finding a conclusion that is science based and accurate using the most current technology.

Questions for Discussion

1. Considering other methods available, can tree-ring dating help you find information about a flooding event (or geographical extent)?

2. What else can tree rings tell us about our local climate? Discuss in class.

3. How can biosolids impact the growth of trees and readings of tree-rings?

4. Ask a lawyer if using tree-ring data to verify atmospheric conditions at a certain date/time would hold up in court. i.e. major flooding, excessive wetting periods/ dry season....

Chapter 11
Automotive Accidents and Impacts from Weather Disasters

Probably the most common request you'll receive will somehow be associated with a car accident (depending on geographical location). Most of the calls I have received concern accidents from 12 months to 5 years ago and include anything winter can throw your way. Let's face it, winter storms are deceptive killers! Since most of the deaths that occur, are indirectly related to the actual storm.

If there was ever an elective science class for law enforcement students and police officers, it should be a meteorology class. Since the weather plays a vital role in accidents, the first on the scene, after assessing for injuries, the officer should take an observation of surface and air conditions. Record air temperature, winds, type of precipitation, accumulation of precipitation on the ground - basically a complete weather report. (A similar report is recorded on notification of an aircraft mishap). Condition of the roadway is also important. Go so far as to measure the amount of ice/packed snow on the road where the beginning stage of the accident started and ended. You can obtain a mobility weather station (hand held) –

from a number of on-line weather companies.

Kestrel 4000 Pocket Weather Tracker from Ambient

Re-occurring certification, bi-annually, on this subject would insure a higher degree of accuracy in reporting the weather on traffic citations.

If there's a disagreement on whether there was ice, check the scene of the accident for any skid marks. If it's been years, you may not have any physical evidence. Check around (conduct interviews) with local businesses owners, entrances where trucks may wait to enter – this may warm the ground where some ice could form after the truck leaves the spot. Work with other investigators and learn what they look for. This will only strengthen your sensitivity to notice minute details which could be a factor.

Edmond Locard's Principle of Exchange states that when any two objects come into contact, there is always transference of material from each object onto the other.
http://www.computing.surrey.ac.uk/ai/impress/ 06-19-2003

- People die in traffic accidents on icy, snow packed roads.

Transportation-related injuries can generate complex and expensive litigation, requiring extensive expert testimony, time-consuming discovery of defendants' paper and sophisticated navigation of the federal, state and local transportation laws and regulations that often control the result in a major transportation case. Check out this site to see a statistical listing of accidents.
http://www.car-accidents.org/stats-conditions.htm

These accidents can vary tremendously and it's up to you, the forensic meteorologist, to get a good idea as to the presumed cause of the accident as well as other factors that have been documented. The deeper you dig for answers, the better your chances of finding other factors that amplified the cause of the accident. Here's an example;

It's a cold, winter day, snow bands have reduced visibility and some parts of the road had minor drifting (couple of inches). Winds were blustery in open fields and the temperature was quite cold, -22 degrees F. Speed limit for the road was posted 55 mph.

Greg MacMaster   Environmental Forensics

The accident report indicates that a driver spun out and hit another car and that there were injuries although none severe – at that time. Two years go by and you're requested to gather data for a lawyer to determine who bad it was snowing and if there was ice on the roadway. This should sound familiar as it was discussed it in a previous chapter. At a glance, with pertinent data available (snowfall records, plow schedule, observation data from the closest station a few miles away) tend to learn towards a typical snowy day with blowing and drifting. Operator error, driving too fast for conditions or other decisions made by the officer, that could show fault.

If you were to type up a report of weather conditions at that time and preceding 6 -12 hours, you would simply refer to the observational data sheet and compare with local plow records to substantiate the claim that there was snow on the ground. A one page expert witness report could easily be typed and sent to the requesting lawyer within a day or two. It's an easy couple hundred dollars, right? But the idea that I'm trying to instill here is to look at all the possible ways the accident could occur (within the realm of meteorological explanation). Tree line, protected areas from wind, inclines declines, intersections or anything else that could cause a change in the apparent wind/weather condition. To see this means you'll have to go to the site and physically check yourself.

Intersections is another place where accidents occur and drivers have little knowledge about something we, as meteorologists, can see very well. Glazed ice! Here's how it forms: It's snowing and air temperatures are well below freezing, so there's no way for rain to form right? Well, we have heat generated from car engines that melt the snow (to varying degrees). As traffic resumes, the melted snow then freezes. This process continues every 3 minutes from when the traffic light turns green to red. You can now see the process of freezing and melting for hours at a time. Glazed ice forming at intersecting roads makes a perfect combination for accidents to occur. When any two objects come into contact, there is always transference of material from each object onto the other. *Edmond Locard's Principle of Exchange.*

This is very evident when car tires carry amounts of salt and calcium chloride (ice melting). This exchange of material will have an affect on what is "traditionally pure water" on the ground. If the chemical composition of water changes, so does the freezing point.

Now on to sensors: Weather sensors are located 2 meters (almost 6 feet) from the earth's surface. It's a constant around the world. Every weather station that is recorded with the WMO, NOAA, NWS are all 2 meters above the ground. If the temperature is recorded to be 33F, 34F or 36F degrees, does that mean there was no ice on the ground? Cold air sinks (heavier) and if you did a study of air temperature verses ground temperature, you just might find a difference as high as 5-7 degrees (depending on mixing due to strength of winds). This could give credit as to the ice on the ground when temperatures were above freezing. Solar radiation and wind flow could offset this finding, so it's a good idea to know and understand the thermodynamic principles and formulas.

Understanding these formulas will give you better insight on how the atmosphere works and is a great way to find information from limited availability of data. Often used in depositions and courtroom testimony, the chart would show conclusive evidence with theoretical and sound reasoning relieving you of the "Junk Science" accusation. Determination of unreported meteorological quantities from plotted soundings (RH%, Vapor pressure, Mixing ratio, Thickness, Pressure and Density Altitude, LFC, LI, SSI, CCL, Equilibrium level and many other values) can easily be definable and used in your interpretation of what elements may have been a factor. To learn more about the Thermodynamic Diagram, plotting and analyzing, write to Cyclogenesis@charter.net and request information on "How to Plot and Analyze a SKEW T log p Diagram". It will come in handy

Tools and supplies needed on cases may vary, but it would help to have the following;

- Camera (digital preferably)
- Tape measure
- Distance measuring equipment (long rope, yardage indicator) to measure distance from trees or other obstacles at a distance

- Temperature sensors (pocket size will suffice), Used to measure 2 meters above the ground, at ground and below ground level.
- Pad of paper (graphic and lined)
- Topographic maps showing area to be surveyed
- Check the slope of the highway from centerline to edge – for drainage (or lack of).
- Make note of objects close to road that would block the sunlight (favorable for formation of black ice)
- Any other objects that seem out of place

As you might conclude, this list may be modified to suit your own personal investigation needs. But to get a good assessment of the location and variables, you'll need a good reference to work from. Supporting evidence showing lack of ice (black ice/packed snow/ liquid) would be the presence of skid marks.

What follows is a detailed examination on speed/braking action and formulas.

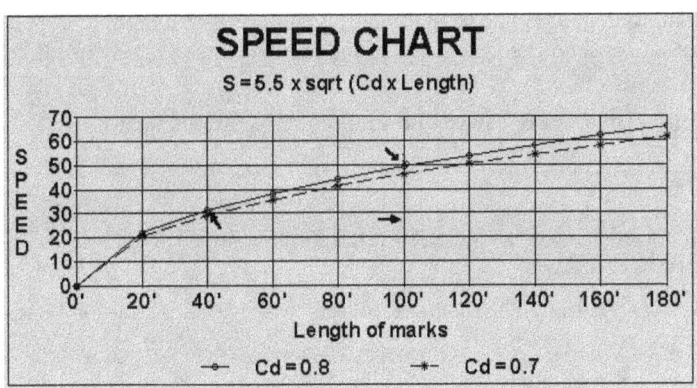

SPEED CHART
$S = 5.5 \times \sqrt{(Cd \times Length)}$

1. Determine the speed in mph after the vehicle has stopped sliding. This might be zero, but generally won't be zero. If the vehicle strikes something at the end of the skid, it is still moving and this residual velocity must be allowed for when using the chart.

2. Move out (horizontally) on a speed line until it hits one of the curves, the lower is for a 0.7 drag factor, the upper is for 0.8. This is the starting point for finding the speed drop associated with a given length of skid marks with a given residual speed after the mark ends.

3. Move out the length of the longest single skid mark, then move up until you hit the curve again to find the original speed.

EXAMPLES:

a. A vehicle leaves 40' of skid marks before coming to rest- the original speed is about 30 mph for both drag factors.

b. A vehicle leaves 60' of skid marks before striking another vehicle, the damage and post impact slide suggest that its speed was 30 mph when it struck the other vehicle.

1. Move out from the 30 mph line until you hit the curve at about 40'.

2. Add the length of the skid mark -60'- go out to the 100' mark and then up until you hit the curve again at about 46-48 mph. This is the original speed of the vehicle.

Marine accidents are different in nature in that most of the accidents that occur are from falling overboard or negligence on the part of the operator. However few the cases involving the weather, you still need access to information that could substantiate the claim of damage. Radar, lightning, satellite and local marine/surface observations are a few of the requests I have had to fulfill.

Time, Distance & Speeds Fundamental Relationships:

1. An easy one: Distance = Speed x Time -Example: 30 mph x 10 hrs = 300 miles

2. A useful one: 1 mph = 5280 ft/hr /3600 sec/hr = 1.47 ft/sec. So: 20 mph = 29.4 ft/sec.

3. **Acceleration:** The **distance** covered while accelerating for T secs.:

$D = V_o \times T + 1/2 \ A \times T \times T$ where: V_o is the vehicle's speed when it starts accelerating (0 for a standing start) and A is the acceleration rate- for cars this is about 10 ft/sec/sec.

Example: $V_o = 0$, T = 2 sec. D = 1/2 x 10 x 2 x 2 = 20 ft.

The **time** required to cover a given distance from a standing start:

$T = \sqrt{2 \times D / A}$ - Many calculators will do square roots.

4. **Braking Distance:** The total distance required to stop the vehicle equals the distance covered while the driver is detecting and reacting

to the situation plus the distance covered while braking. So $D = V_o \times T$ + the braking distance :

$Db = S \times S /(30 \times Cd)$ where:

S is the speed in mph and Cd is the drag or braking coefficient, 0.7 - 0.8.

Braking time: Take the speed in ft/sec and devide it by the deceleration rate- about $0.7 - 0.8 \times 32.2 = 22 - 26$ ft/sec/sec. Add the reaction time.

5. Speed from skid marks:

If the vehicle is stopped at the end of the skid marks, then the original speed is : $S = 5.5 \times \sqrt{(Cd \times \text{Length of skid marks})}$ where the Cd is as defined above and the speed is in mph. If the vehicle is still moving when the skid marks end, this formula does not yield the original speed. (see the "Combined Speed" sheet)

6. Speed from damage:

It is possible to compute speed changes from crush damage but the calculations are complicated and generally require a computer along with damage measurements and knowledge of the structural properties of the vehicles involved.

The Combined Speed Formula

$$S^2 = S_1^2 + S_2^2 + S_3^2$$

- With a multiple phase accident where a vehicle may have undergone several speed changing events, it is necessary to use the combined speed formula (above) to estimate the original speed of the vehicle.

- If for example, a vehicle looses 30 mph in sliding before impact, 20 mph in the collision and 10 mph in sliding to a stop after the impact, the original speed of the vehicle is then:

$S \times S = 30 \times 30 + 20 \times 20 + 10 \times 10 = 900 + 400 + 100 = 1400$, so that $S = 37$ mph

Notice that we can't simply add the speed decrements together, they must be combined using the formula. This frequently leads to lower speed estimates than might be anticipated when looking at the various speed numbers before they are combined. The reason is that speed energy- what's really being computed in the partial estimates- varies with the square of the speed. Thus 20 mph is four times the (kinetic) energy of 10 mph (20 x 20 =400, 10 x 10 =100), not twice as much. The speed from skid marks formula - S = 5.5 x sqrt (cd x distance) - is a computational aid which assumes a zero final velocity, that is, that the vehicle has stopped at the end of the skid marks; if the computed value is to be taken as the original speed of the vehicle.. It does not yield speed drops simpliciter. Thus, 20 feet of locked wheel skid marks represents about 20 mph if the vehicle is stopped at the end of the marks. Suppose however that the vehicle is doing 60 mph when it leaves 20 feet of skid marks. Its speed at this point is: S = sqrt (3600 - 400) = sqrt 3200 = 56.6 mph.

Weather Impacts in Canada

(Environmental Adaptation Research Group, Atmospheric Environment Services Downsview, Ontario, Canada)

Canada is subject to a variety of hydro-meteorological hazards and disasters. An inventory is being assembled and Figure 1 shows the number of identified events by hazard for which cost estimates could and could not be made. The total of known costs is identified in 1995 dollars at the end of each bar. Tornadoes are not included, since there have been over 2,300 known Canadian tornadoes during the period 1961-1996. I would also note that a precise criteria was not used to include or exclude particular events but rather events were included if the information source suggested a "significant" impact of some sort (meteorological, social or economic).

Figure 1 - No. of Identified Events

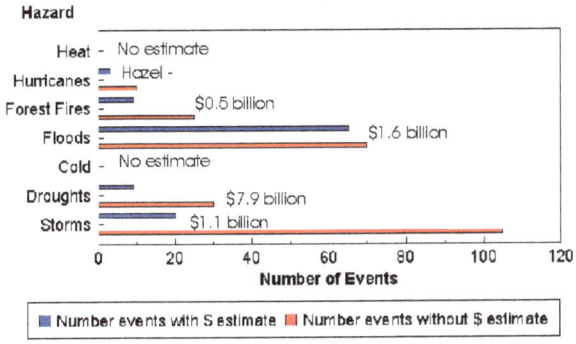

Droughts, by far, are the most costly hazard, although they rank fourth in terms of frequency. It is worth noting that for hurricanes, only economic costs associated with Hazel (1954) have been included. An historical survey of Canadian disasters shows that 44% of them are weather or climate related, that almost 1/3 of all disaster has occurred at sea, and that 80% of those were weather related.

Natural disasters are the extreme of natural hazards and occur when social vulnerability is triggered by an extreme event. The costs Canadians incur from hazards are a function of our adaptive decisions. Unsafe conditions result from a number of social forces which are rooted in limited access to power, economic resources and the nature of political and economic decisions.

These costs of natural hazards can be broken down into social costs and economic costs.

Social Costs

In terms of social costs the following statistics give some indication of the overall magnitude:

Transportation

Decreasing trend of weather-related aircraft accidents from approximately 60/year in 1985 to less than 20/year since 1992.

Although there have been few fatalities, there has been a significant number of weather-related railway accidents averaging between 20-40 per yea; however, 100-120 occurred in 1990 and 1991.

Historically, many of Canada's worst disasters have been ocean based with typically 300-400 weather-related marine accidents each year (note 1990 worst year in the past 10 years with slightly less than 500 weather related marine incidents).

Major road accidents result mainly from wet conditions, followed by ice, snow, slush and mud. In 1992 weather-related road accidents resulted in 298 fatalities, over 23,000 personal injuries and over 72,500 property damage incidents.

Deaths

Canadians occasionally die as a result of atmospheric hazards. Most deaths occur as a result of cold. In the past decade, however, the number of deaths from cold has shown a gradual decrease, while those resulting from other atmospheric causes have remained fairly constant (note: some concern as to how attribution of cause has been assigned).

Economic Costs

There are two fundamental economic costs associated with natural hazards - adaptive costs and impacts, response and recovery costs. Adaptive costs are those associated with protection, reduction of vulnerability or risk, education and research. These costs are difficult to estimate and little research has been devoted to increasing our understanding of these costs. One preliminary estimation of Canadian adaptation costs suggests that $13.7 billion is spent annually; however, this is likely underestimated by some unknown amount.

Impacts, response and recovery costs are those incurred when protection fails or no adaptive response is taken. Some examples for Canada include:

Forest Fires can have a direct impact due to the loss of natural resource, though it is unclear how to account for these losses as fires are now considered an essential part of the natural ecological cycle. The annual area burnt suggests an upward trend with 1995 being by far the worst year with over 7 million hectares burned, followed by 1989. This statistic is related to climate, but is also related to decisions made regarding fire management. All provinces incur costs related to fire management. Over the last decade, Ontario spent over $800 million, more than any other province.

Annual fire management costs are shown on figure 2 with costs peaking in 1995 at over $450 million.

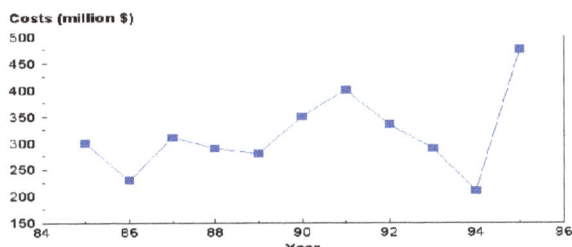

Figure 2 - **Fire Management Costs**

Hydro Companies

Weather-related costs are highly variable as indicated by those for Ontario Hydro which has annual weather-related costs ranging from zero to $3 million and averaging $1.4 million/year.

The most costly hazard was "wet snow + high winds" the total cost which occurred due to one event at Vegreville and Lloydminster, Alberta. This two-day storm destroyed 108 steel transmission towers, 300 wood transmission structures, and more than 3000 wood distribution poles. In addition more than 400 km of conductor had to be replaced.

Tornadoes (numbering 8) come second in terms of costs, though they were all from one utility company.

Federal Payments

Floods have cost the most, by far, with audited totals over $300 million and federal payments of $150 million (does not include the 1996 Saguenay flood).

Storms and fire rank second and third with about 11% each.

Provincial Costs

Provinces incur costs due to crop damage from hail, flood, drought and a variety of other hazards.

The largest annual expenditure occurred in Saskatchewan, with payments totally over $500 million (1988 and 1989).

Drought is the major hazard resulting in crop insurance payments on the Prairies (e.g., Manitoba paid out $400 million from 1966-1994 in response to drought-related crop losses, followed by excess moisture, hail, heat and frost).

Insured Costs

Figure 3 shows the costs of multiple major payments from 1983-1994 (does not include the costs of events less than about $4 million, and therefore the true costs are much greater than those shown in this figure).

Hail has caused the most damage (over $450 million), followed by tornadoes, flood, storms and wind.

There appear to have been 9 events in 1995 which include significant damage from flood, hail, thunderstorms, wind and Hurricane Hortense.

The largest single disaster in Canada was the Calgary hailstorm of 1991, which cost insurance companies around $400 million. The insured costs of the Saguenay floods are currently estimated at $350-400 million.

Figure 3 - **Weather Related Insurance Costs**
Major Multiple Payouts

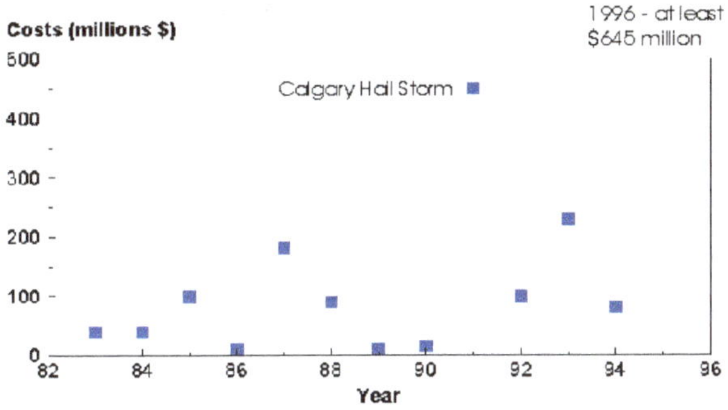

1996 - at least $645 million

Canadian Weather-Related Disasters in 1996

In 1996, Canadians suffered through some of the most extreme and destructive weather ever to have hit the country. For most of the year, the weather either froze, buried, soaked, buffeted or frightened Canadians. No part of the nation seemed to escape the wrath of the weather in 1996.

Total property damage will likely exceed $2.5 billion when the final figures are tallied. Indirect costs and losses from revenue shortfalls, canceled events, missed opportunities, and slowed business will probably be in excess of a $3 billion hit to the Canadian economy. Remarkably though, the number of personal injuries and fatalities linked to weather incidents could have been much higher. Unofficial numbers point to fewer than 25 weather-related deaths (excluding deaths from road accidents and hypothermia) - 10 from the storm in the Saguenay and six from lightning in separate incidents. The top weather stories of 1996 are:

The Saguenay Flood

By far the worst catastrophe of the year, and Canada's first billion dollar plus natural disaster, was the flooding and mud slides in Quebec's Saguenay River valley in mid-July. The storm produced the largest ever overland deluge in Canada this century - an amount equivalent to a two-month flow over Niagara Falls - triggering a surge of water, rocks, trees and mud that killed ten people and forced 12,000 residents to flee their homes. It was the deadliest flood since Hurricane Hazel in Toronto in 1954.

The scale of the tragedy was staggering. Many of the region's roads and bridges and delivery systems for power and water simply disappeared. To the insurance industry it was Canada's worst-ever weather disaster in economic losses. By including insured and uninsured losses and indirect costs to the economy, total losses are sure to exceed $2 billion.

The Pacific Storm of 1996

During the period December 22 to January 3, 1997 a series of brutal winter storms blasted Vancouver Island, the Lower B.C. mainland and the Fraser Valley with up to 85 cm of snow paralyzing infrastructure and commerce. The accumulation of snow for this series of storms was unprecedented in historical records of Vancouver Island and the lower south coast of British Columbia. The recorded 64.5 cm snow that fell in a 24 hour period, at the Victoria airport, gave Victoria the distinction of having the 3rd highest snowfall, in a major city, in Canadian history. During the same period, Downtown Victoria received 85 cm.

The environmental impact was severe in that massive releases of untreated of partly treated effluent were released into rivers and the ocean.

Although estimates of the economic losses range in the area of $200 million, the true impact of the storm will probably never be known. The following list suggests the overall impact

- over $200 million in total losses estimated
- largest search and rescue team in the history of B.C. dispatched
- 170 motorists stranded and later rescued
- 1700 people received emergency food and shelter
- Trans Canada highway closed for two days –
- stranding   thousands
- secondary arteries closed for up to 10 days

- police and fire departments unable to respond
- 150,000 homes without power
- Roofs/buildings collapsed under heavy snows
- over 500 avalanches reported
- 65 reports of pollution spills
- hundreds of flights canceled
- ski resorts closed
- theaters and sporting events canceled
- food shortage

High Energy Costs

In much of Canada, 1996 featured one of the longest and most vicious winters in recent memory. Three straight weeks of frigid weather gripped almost the entire country in January making it colder in most cities in western and central Canada than it was at the North Pole. To keep up with the cold, utility companies pumped out power in record amounts from British Columbia to New Brunswick. Canadians paid an additional $500 million to keep their dwellings as comfortable as in the previous winter.

Costly Prairie Hailstorms

In July, hailstones the size of fists bombarded Winnipeg and Calgary, racking up close to $300 million in property losses. In Manitoba, more than half the losses were for auto damage, making it the worst single disaster claim against the Manitoba Public Insurance Corporation in its 25-year history. At least a third of the cars damaged had to be written off. In Calgary, hail and flooding rains knocked out the city's 911 service and swept away cars.

Wet and Cold Weather Reducing Crop Yields

Unfortunately for western farmers, prospects in early September for one of the most bountiful grain crops in Canadian history didn't exactly materialize. Fall temperatures across the west were much below normal (the Prairies had their second coldest fall in half a century) and precipitation was much above normal (the 7th wettest fall in half a century). Cool wet weather during the harvest of western red spring wheat led to a severe drop in its grade distribution, denying farmers an additional $180 million.

In southern Ontario, winter wheat production was severely affected by the wet cool weather throughout the growing season. Record rainfalls resulted in the worst outbreak of blight fungus ever seen in Ontario. According to Agriculture Canada, the excessive moisture and disease not only reduced yields, but it also reduced the quality of most of the crop to feed, since affected grain cannot be used for human consumption. The loss was estimated to be about $90 million.

Deep Winter Snows

So much snow fell early in the winter that before 1996 even started, many cities in western and central Canada had all but exhausted their snow removal budgets. Hardest hit was the central Ontario snowbelt from Barrie to Sault Ste. Marie, where on several occasions, cars disappeared in snow drifts, service centers became refugee camps, roofs collapsed and schools closed.

Insurance claims paid in the first three months of 1996 were 11% higher than in the previous year when the weather was much less severe. Total insured losses owing to the weather estimated at $165 million.

Slow Spring Affecting Retail Sales

For most of Canada, the winter season gave way to the monsoon season. Unrelenting rains and dreary weather plagued the country from April to June. Garden centers and golf courses were virtually empty during the spring. Retailers blamed the persistent cool and rainy weather for a 30% drop in the sales of weather sensitive goods and services, such as pools, air conditioners and warm season apparel. Sales of general merchandise in April and June were down by $100 million over the previous year's sales.

Flash Flooding in Ottawa and Montreal

The third major storm in less than two weeks, and the worst on record, hit Ottawa-Hull in early August with a deluge of 100 to 150 mm of rain in 90 minutes. Total insured property damage exceeded $20 million, not including the costs of repairs to damaged sewers and roads. Between November 7 and 9, thirty hours of steady rains drenched parts of Montreal and southwestern Quebec. The rains washed out sections of highways, collapsed bridges, derailed trains and undermined road and rail beds. Damage estimates put the event at over $60 million.

Severe Thunderstorms and Tornadoes

The snow had hardly melted in southern Ontario when the season's first tornadoes tore through regions east of Lake Huron in April. The twisters injured two people and caused property losses, much of it uninsured, approaching $8 million.

Severe thunderstorms on July 4 spawned at least eight tornadoes in Saskatchewan. Winds of 140 km/h and hail the size of golf balls produced $15 million in property damage. Two weeks later seven tornadoes touched down in Alberta, trashing trailers and flattening granaries to the tune of $10 million. Near Stoney Plain, more than 100 mm of rain fell in severe thunderstorms backing up sewers and flooding basements for another $10 million in losses. Tornado-related damage in Canada easily exceeded $50 million.

Spring Flooding

Significant flooding occurred in several communities across Canada during much of the spring and early summer. The Okanagan experienced its worst flooding in six years, The Red River inundated farm fields, roads and major highways leading authorities to declare a provincial flood disaster for the first time since 1979. In Winnipeg, the costs of filling 336,000 sandbags and protecting pumping stations alone cost $1.2 million. In Timmins, Ontario the Mattagami River overflowed its banks in the worst flooding across Canada and losses are still being tallied, with final figures are expected to range between $20 and $50 million.

Hurricanes and Weather Bombs

Four hurricane-force storms struck Eastern Canada in 1996: Bertha, Edouard, Fran and Hortense. It was the second consecutive season with above average hurricane formation in the North Atlantic. In 1996 there were 13 named storms of which 9 were hurricanes, including six intense ones, compared to a normal of 9 storms, 6 hurricanes and 2 intense ones.

Hortense, which swept east of Halifax and traversed western Newfoundland on September 14, was the first hurricane to achieve landfall in Canada in 21 years. Winds topped 161 km/h on cape

Breton Island, felling trees, lifting roofs and blowing out windows. Total property losses approached $5 million.

Described as the worst storm since Typhoon Freda in 1962, a "weather bomb" struck Vancouver Island on October 17 causing massive power outages while felling trees, setting adrift 50 pleasure boats and ripping apart docks. A "weather bomb" is a storm which intensifies very quickly and moves faster than a hurricane. This storm packed winds as strong as 161 km/h and produced waves as high as 30 meters.

Conclusions

Are the number of weather-related disasters in Canada increasing? According to the Insurance Bureau of Canada, ever since the Edmonton Tornado of 1987, the number of multi-million dollar losses from weather disasters has been on the rise in Canada (note: before 1987 there was no single natural disaster with damages exceeding $1 billion anywhere in the world, let alone in Canada).

While the outburst of extreme weather was interesting, climatologists were not generally surprised by them. Climatologist are, however, becoming increasingly concerned that the volatile weather in 1996 might be a dry run of extreme conditions we might expect from a warming climate.

With this in mind, planners and decision-makers should note that natural hazards and disasters are expensive but not inevitable. With appropriate planning to reduce vulnerability, their social and economic impacts on Canadians can be reduced.

Retail Sales

Why do retailers continue to blame the weather for poor retail sales?

As retailers post annual results, the UK's weather gets the blame once again. Retailers cannot continue blaming the weather for poor business performance they must begin managing weather related risks. The correct weather forecasting products and services could soon make the 'wrong weather-poor performance' excuse invalid.

The impact of weather on retailers: Research by Datamonitor states that UK grocery and clothing retailers could be missing out on as much as £4.5 billion per year by failing to include weather forecast information in their short and long term business planning decisions. There are numerous examples of unpredicted weather hitting bottom line retail results.

House of Fraser was reported in Retail Week, June 2003, as stating: 'According to analysts, the Iraq conflict, followed by poor weather in May contributed to a tough trading climate on the high street.'

A sales and trading update from Thorntons, specialty retailer and manufacturer of high quality chocolate, stated: 'The exceptionally hot weather significantly reduced sales in the week leading up to Easter, although all egg stocks have been sold at discounted prices.' (Trading Statement by Thorntons, May 2003).

In April 2003, BBC News reported: 'Scotland's retail sector has dramatically outperformed the rest of the UK. Unseasonably warm weather in Scotland helped to lure Scots out for retail therapy.'

In October 2002, BBC News also reported: 'Next and Ted Baker have reported a surprise slowdown in sales, blaming unusually warm autumn weather. Both companies stated the Indian summer sunshine had deterred shoppers from buying their autumn and winter ranges.'

However, the UK's weather is not always a problem. For some retailers, the weather has provided significant gains. Burberry for instance, reported that the cold and wet weather over Christmas 2002 helped to drive trade in hats, scarves and coats (BBC News, Jan 2003).

Fashion retailer H&M, also benefited. In November 2002, BBC News reported: 'Sales jumped by 26% in October compared with last year. Like most fashion stores, H&M suffered due to unusually warm weather in September, which meant that people were not buying autumn clothes. However, a cold weather spell in October had helped to boost demand.'

Classic examples of UK weather impacting product sales include ice cream, chocolate, de-icer and umbrellas, which are clearly weather sensitive. Other examples are less obvious. For instance, each summer, UK supermarkets sell more toilet rolls than at any other time of year – around 20,000 more packs each month. The reason is that people suffer from hay fever during the summer and use significantly more tissues.

Managing weather related risks: Most retailers see improving supply chain and business-response lead times as ongoing goals; the next step is to make quick process changes in response to weather conditions.

More retailers are considering insurance to cover weather risks. It is possible to trade in weather derivatives to provide financial protection from unpredicted losses due to weather. However, purchasing weather related insurance is a new concept and few retailers have the knowledge, skill and experience to select the correct options. Some are following the direction taken by some of the worlds' leading retailers and adding weather forecasting and consumer demand models to the business decisions process.

Accurate weather information could improve the timing of key phases such as the introduction of high-summer fashion clothing. Allocations and store replenishment could benefit from moving forward the production, intake and allocations of summer clothing. Special promotions and advertising could be closely linked to weather forecast information and predictions.

What are the benefits for retailers that use weather forecasting? The benefits depend on how vulnerable products and services are to the weather, and whether forecasting information can be applied rapidly enough to allow retailers to make material changes to their business. However, it would be reasonable to expect the benefits to include increased sales, increased product availability, reduced markdowns and increased margins.

Retail Week recently stated that Safeway acted on accurate forecasts for hot weather in the run up to Easter 2003 and successfully modified their in-store promotions to deliver an extra £2million of sales.

What are the options? There are a number of organizations offering services for short, medium and long-term weather forecasts. They analyze the relationship between product sales and weather elements, and predict future consumer demand. These solution providers vary from institutions such as the Meteorological Office, through companies such as Planalytics in the USA, to small UK-based partnerships such as Weather Commerce and Smart Research.

Making the most of weather forecasting techniques. Retailers must first understand their business requirements to use weather forecasting techniques to improve bottom line results. This means recognizing how their inability to predict the weather has impacted the business. It also demands an understanding of the processes and systems that require improvement.

It also needs a detailed knowledge of weather dependant product ranges and how they will be affected. Timing is also critical. Could retailers make business improvements based on weather forecasts up to 10 days, 30, 90 or over 90 days? There will be an optimum window and the trick is to identify it.

Weather related forecasts can improve business planning, supply chain and merchandising decisions. The potential benefits of managing weather related risks could offer significant return on investment. In the future, shareholders may begin to lose patience with retailers who continue to cite the vagaries of the weather as an excuse for poor performance.

Retail Sales (Courtesy: Paul Mason Consulting. RFID)

Questions for Discussion

1. In relation to the Speed & Breaking Action chart, if icy patches were present – or proven to be present at time of accident, could this alter the assumed speed of the vehicle? Explain.

2. You receive a call from a business owner asking for weather data over the course of several months, what do you ask before providing this information?

3. What business could minimize loss in retail sales (by knowing the long range forecast) when ordering seasonal products?

4. How popular is the Farmers Almanac in your region? How accurate is it?

Chapter 12

Drowning Victims

While much of this chapter is designed for the recovery expert and identification of body in various stages of decomposition, finding the body is of vital importance (see the chapter on Body Recovery and the Forensic meteorologist – a Case Study). The Forensic Meteorologist can drastically improve the chances of finding a submerged body by using past wind and adjusting for hydrodynamic undertow.

Immersion artifacts occur in any corpse immersed in water, irrespective of whether death was from drowning or the person was dead on entering the water.

Therefore, immersion artifacts do not contribute to proof of death by drowning. However, such artifacts are typically the most striking findings in a body recovered from water. These immersion artifacts include:

1. goose-skin, or anserine cutis, which is roughening, or pimpling of the skin,
2. skin maceration, or washer-woman's skin, which is swelling and wrinkling of the skin, and
3. adipocere, which is the transformation of the fatty layer beneath the skin into a soap-like material - a process requiring many weeks or months.

Corpses in water always lie with the face down and with the head hanging. Buffeting in the water commonly produces post-mortem head injuries, which may be difficult to distinguish from injuries sustained during life. The presence of bleeding usually distinguishes ante-mortem from post-mortem injuries. However, the head down position of a floating corpse causes passive congestion of the head with blood, so that post-mortem injuries tend to bleed, creating the diagnostic confusion.

The normal changes of decomposition of a body are delayed in cold, deep water so that bodies may be surprisingly well preserved after a long period of immersion. These conditions also favor the formation of adipocere which protects against decomposition.

When a body is recovered from water, two critical questions require resolution: Was the victim alive or dead when he entered the water? Is the cause of death drowning? (and if not, what is the cause of death?).

To resolve the above questions, the following information must be correlated:
1. the circumstances preceding the death,
2. the circumstances of recovery of the body, and
3. the autopsy findings. The approach should be to consider the circumstances revealed by the investigation and to then determine if the autopsy findings are consistent with those circumstances.

Drowning is "suffocation due to immersion of the nostrils and mouth in a liquid". The mechanism of death is complex and is not simply asphyxiation due to suffocation.

Submersion is followed by struggle which subsides with exhaustion and drowning begins. When the breath can be held no longer, water is inhaled, with associated coughing and vomiting, and is rapidly followed by loss of consciousness with death some minutes later.

Instantaneous death may occur following sudden, unexpected immersion in cold water. This is "atypical drowning" due to vagal inhibition - a sudden stopping of the heart mediated by the nervous system.

Hypothermia (death from loss of body heat) may occur following immersion in water with a temperature less than 68oF. A healthy person in ordinary clothes and wearing a lifejacket would have an expected survival time of less than three-quarters of an hour at temperatures less than35oF, less than 11/2 hours at 35-40F, and less than 3 hours at 40- 60F.

There are no autopsy findings pathognomonic of drowning. Consequently, obtaining proof that the victim was alive on entering the water and excluding the presence of natural, traumatic and toxicological causes of death, are critically important. Some pathological changes are characteristic of drowning, but the diagnosis is largely one of exclusion.

A fine, white, froth or foam in the airways and exuding from the mouth and nostrils is characteristic of drowning. It is a vital phenomenon and indicates that the victim was alive at the time of submersion. However, similar foam is found in deaths from other causes, e.g. heart failure, drug overdose, and head injury.

The lungs are characteristically over-inflated and heavy with fluid. However, this is not invariable and, when present, is not distinguishable from "fluid on the lungs" (pulmonary edema seen in heart failure, drug overdose and head injury). It is disputed whether sand, silt, weed, and other foreign matter, found in the airways constitutes proof of immersion during life. The presence of large quantities of water and debris in the stomach strongly suggests immersion during life. Conversely, the absence of water in the stomach suggests either rapid death by drowning, or death prior to submersion.

Hemorrhages in the boney middle ears are occasionally (some would say commonly) seen in drowning cases. Such hemorrhages also occur in deaths from other causes, e.g. head trauma, electrocution and mechanical asphyxiation. There are no universally accepted diagnostic laboratory tests for drowning. The diatom testis used in some British laboratories and may provide corroborative evidence of death by drowning.

INVESTIGATION

These cases represent a challenge because:
1. The mechanism of death in drowning is neither simple nor uniform.
2. The circumstances of drowning introduce further variables.

The questions to be resolved by the investigation are:
1. Did death occur prior to or after entry into the water? (i.e. was the victim alive or dead at the time of entry into the water?)
2. Is the cause of death drowning? If not what is the cause of death?
3. Why did the victim enter the water?
4. Why was the victim unable to survive in the water?

To resolve these issues the following information must be correlated;
1. Circumstances preceding the death.
2. Circumstances of recovery of the body from water.
3. Autopsy and laboratory analyses.

A full investigation of the circumstances preceding the death requires the identification of the victim which therefore becomes a priority. The correct interpretation of the autopsy findings and indeed the performance of some autopsy procedures is dependent upon a careful examination of the circumstances preceding death and of body recovery. The approach should be to consider the circumstances revealed by thorough police investigation and to then determine if the autopsy findings are consistent with those circumstances. In those instances where the cause of death is drowning it is then necessary to answer the questions 3 and 4 above. The investigative scheme must encompass both environmental factors and human factors. "Human factors drowning" implies that human deficiencies were the significant factor in the drowning episode e.g. inexperience, poor judgment, intoxication.(modified from Davis 1986)

Investigative Considerations - Case Example:

An 80 year old male swimming in warm ocean water in mid-summer. He was found dead in the water and the body recovered. Autopsy disclosed minor degrees of pulmonary congestion andoedema. There was severe coronary artery atherosclerosis with posterior wall myocardial fibrosis but no evidence of recent infarction or coronary thrombosis. Consider the environmental and human factors.

Some alternatives are:
1. A fatal cardiac dysrhythmia with a collapse "dead" into the water.

2. A fatal cardiac dysrhythmia with collapse into the water and agonal aspiration of some water.

3. A non-fatal cardiac dysrhythmia with syncope and collapse into the water and drowning.

4. Stepping into or being swept into deep water and an inability to escape due to a lack of cardiac reserve or lack of cardiac rhythm stability.
5. Stepping into or being swept into deep water, panicking and drowning while the heart continued to function normally until overcome by the terminal anoxia of drowning.

Given these alternatives the death certificate might read:

1. Atherosclerotic coronary artery disease
2. Atherosclerotic coronary artery disease with a contributory effect of "agonal aspiration of water".
3. Atherosclerotic coronary artery disease with a contributory effect of drowning.
4. Drowning with a contributory effect of atherosclerotic coronary artery disease
5. Drowning.(modified from Davis 1986)

CIRCUMSTANCE AND MANNER OF DEATH
The world incidence of death by drowning is estimated at about 5.6 per 100,000 of population. Approximately 1,500 deaths from drowning occur in the UK each year; 25% occur in the sea and the rest in inland waters; the majority of victims are young adults and children; two-third are accidental and one-third are suicidal; homicide by drowning is rare.

Accidental drowning of toddlers may be in uncovered fish ponds, the bath and swimming pools. Accidental drowning in adults is commonly associated with alcohol consumption and males predominate. In suicidal drowning some clothing may be left in a neat pile close to the water.

The pockets may be filled with stones or weights may be tied to the body. The hands or the feet are sometimes tied together and an examination of the ligatures will show whether they could have been tied by the deceased. There may be concurrent use of other suicide methods such as drug overdose or slashing of the wrists; alternatively there may be a history or autopsy evidence of previous suicide attempts. Persons jumping from a bridge or cliff into water may suffer injuries from impact with rocks or the water itself.

Impact with the water can produce severe fatal injuries such as fractures of the ribs, sternum and thoracic spine and lacerations of the heart and lungs. Homicidal drowning is uncommon and requires either physical disparity between the assailant and the victim or a victim incapacitated by disease, drink or drugs, or taken by surprise. Disposal in water may be attempted where the victim has already been killed by other means. A victim of infanticide is sometimes disposed of in this way. Autopsy is directed towards establishing injuries inconsistent with accident in the absence of signs of drowning. The investigation of a death in a domestic bath may be made more difficult by the lack of accurate information concerning the position of the body as found and the level of the water.

First it must be established whether the nose and mouth were truly under the water. Such drownings will only occur if unconsciousness is produced by disease (epilepsy, coronary arterial sclerosis) or the consumption of alcohol and/or drugs or a head injury from a fall. Suicide in the bath is rare but well documented. Homicide in the bath is described (the "Brides in the Bath"). Intravenous drug abusers may place an individual who collapsed from an adverse reaction to a drug in a water-filled bath during attempted resuscitation; drug paraphernalia are typically found nearby.

Where the victim is a woman of child bearing age then pregnancy and abortion should be suspected. The domestic bathroom presents other hazards than drowning such as electrocution and carbon monoxide poisoning from faulty heaters. Persons unconscious by reason of natural disease and injury can drown in quite shallow water so long as it is sufficiently deep to cover the nose and mouth.

Diving into shallow water may result in impact of the forehead against the bottom with result of a hyperextension of the head and loss of consciousness. Common autopsy findings are hemorrhage in the deep muscles of the neck with or without associated fracture of the cervical vertebrae. Bruises and abrasions on the face or forehead may provide evidence of the impact.

Individuals engaged in underwater swimming competitions may hyperventilate prior to entering the water. This can result in sudden loss of consciousness and drowning. The postulated mechanism is that over breathing so reduces the carbon dioxide content of the blood that there is no activation of the respiratory centre even when the arterial oxygen tension falls to a critical level and consciousness is lost. In skin diving a mask and fins are used and it is essentially an extension of swimming with similar hazards. SCUBA is an acronym for self contained underwater breathing apparatus.

This apparatus allows the diver to reach depths not usually attained by skin divers. The hazards are those of drowning and baro-trauma. The commonest problems include "the bends" (caisson disease, decompression sickness), acute pulmonary emphysema, pneumothorac and systemic air embolism. The latter three are stages of the same process resulting from excessive pressure inthe lungs ("extra-alveolar air syndrome"). Investigation of these deaths requires examination of the scuba apparatus by an expert as well as specialist advice as to the circumstances. Prolonged immersion in water less than 68F carries the threat of hypothermia.

For a person in good health ordinarily clothed and wearing a life jacket, the expected survival time for given water temperatures are:

less than 3/4 hour at less than 35°F;
less than 11/2 hours at 35-40°F;
less than 3 hours at 40-50°F;
less than 6 hours at 50-60°F.

DROWNING Definition: - suffocation due to immersion of the nostrils and mouth in a liquid.

Qualifications: the mechanism of death is complex and varies somewhat with circumstances.

* It is not simply an asphyxiation due to suffocation in a liquid.
* immersion of the nostrils and mouth is the minimal requirement,
* typically the entire body is submerged in the liquid.
* the liquid is most commonly water but drowning can occur in any liquid e.g. beer, wine, gasoline, bitumen, dye, paint or some other chemical solution.

Mechanism of Death Drowning was originally conceived as suffocation due to the mechanical obstruction of the airways by liquid.

The animal experiments of Swann and his colleagues during 1947-51 highlighted the pathophysiological importance of disturbances of blood electrolytes and fluid balance. In the experiments dogs were completely submerged in salt water and fresh water. In fresh water and brackish water (approximately 0.5% salinity) drowning the aspirated water is rapidly absorbed from the alveoli into the circulation producing an expansion of blood volume, hemodilution and hemolysis.

Within three minutes of submersion hemodilution was up to 72%. Circulatory overload, hyponetremia and sodium/potassium imbalance together with myocardial hypoxia resulted in a dramatic collapse of systolic pressure quickly followed in the majority of cases by ventricular fibrillation. In salt water (3-4% salinity) drowning the aspiration of water results in withdrawal of water from the pulmonary circulation into the alveolar spaces as a result of the osmotic differential while at the same time electrolytes (sodium, chloride, magnesium) pass into the blood. There is hemo-concentration but not hemolysis and little change in the sodium/potassium balance. The pulse pressure decreases slowly and is followed by A-V dissociation but not ventricular fibrillation.

Up to 42% of the water content of the circulating blood was absorbed into the alveoli. In both fresh water and salt water drowning there is terminal pulmonary edema. In both drowning media there is concurrent transfer of water in both directions between the alveolar spaces and the blood i.e. pulmonary edema develops simultaneously with the diffusion process.

These experiments have been extrapolated to man but have been criticized because

(a) the animals were always completely submerged and

(b) the main intracellular cation in the dogerythrocyte is not potassium but sodium.

The biochemical findings in humans surviving drowning are less distinct.

Phases of Drowning

1. Submersion is followed by struggle which subsides with exhaustion and drowning begins.

2. Breath holding lasts until carbon dioxide accumulation stimulates respiration resulting in inhalation of water.

3. Gulping of water coughing and vomiting is rapidly followed by loss of consciousness.

4. Profound unconsciousness and convulsions are associated with involuntary respiratory movements and the aspiration of water. Respiratory failure proceeds heart failure in one-third of cases it is coincident in one-third and follows it in the other third.

5. Death occurs within 2 to 3 minutes (see below for "Instantaneous Deaths").

Death is almost invariable when the period of submersion exceeds 10 minutes. The survival rate from potentially fatal salt water submersion is about 80% whereas in fresh water it is less than 50%.

TYPICAL PATHOLOGICAL FINDINGS
There are no pathological findings pathognomonic of drowning. Consequently obtaining proof that the victim was alive on entering the water and excluding natural, traumatic and toxicological causes of death are critically important. Some pathological changes are characteristic of drowning but the diagnosis is largely one of exclusion.
1. Foam in the airways
2. Externally a fine white froth or foam is seen exuding from the mouth and nostrils.

3. The froth is sometimes tinged with blood producing a pinkish color.

If the foam is wiped away then pressure on the chest wall will cause more to exude from the nostrils and mouth. It is persistent and resists submersion for several days (up to a week in winter). The foam is also found in the trachea and main bronchi. The foam is a mixture of water, air, mucus and possibly surfactant whipped up by respiratory efforts. Thus it is a vital phenomenon and indicates that the victim was alive at the time of submersion. A similar foam is found with severe pulmonary edema from any cause such as drug overdose, congestive cardiac failure and head injuries.

2.Emphysema aquosum ("emphyseme hydroaerique")The lungs are voluminous/bulky/ballooned. The pleural surface has a marble dappearance with grey-blue to dark red areas interspersed with pink and yellow-grey zones of more aerated tissue. They feel doughy and pit on pressure. On sectioning there is a flow of watery

material. The appearances reflect active inspiration of air and water and cannot be reproduced by the passive flooding of the lungs with water. However the appearances are not generally distinguishable from pulmonary edema. Contrary to expectations (see Mechanisms of Drowning) lung weights in freshwater drowning are not statistically different from lung weights in salt water drowning.

The average lung weight is approximately 700gm with a standard deviation of approximately 200gm so that in a minority of cases the lungs are"dry". Subpleural petechiae are rare but larger ecchymoses are sometimes seen most often in the interlobar surfaces of the lower lobes. Sub pleural bulla which maybe hemorrhagic are occasionally found. Hemorrhages are the result of tears in the alveolar walls and this is the explanation for the occasional blood tinging of foam in the airways. For detailed histological studies one central and one peripheral section from each lobe is recommended. The tissue should be cut with a sharp knife avoiding squeezing out of the fluid content. The microscopic appearance varies from being suggestive of drowning to entirely normal.

Aspiration of large quantities of water results in over distension of the pulmonary alveoli (emphysema aquosum) the alveolar septum are thinned and stretched with narrowing and compression of the capillaries. The appearances resemble pulmonary emphysema. The intensity of the changes reflects the circumstances of the drowning and are most pronounced in persons who drown over a relatively long period of time coming to the surface several times to inhale air. Foreign material in airways, lungs and stomach, sand, silt, weed or other foreign matter may be found in the airways, lungs, stomach and duodenum of bodies recovered from water. Disputed is whether the presence of such material constitutes proof of immersion during life.

When a victim is dead at the time of submersion, water and contaminating debris may enter the pharynx, trachea and larger airways; small quantities may enter the esophagus and stomach. However water will not reach the terminal bronchioles and alveoli to any significant extent so that the finding of abundant foreign material generally distributed within the alveoli provides strong evidence of immersion during life so long as the body is recovered early (within 24 hours)from shallow water (less than 9 feet deep).

Similarly the presence of large quantities of water and contaminating debris in the stomach strongly suggests immersion during life (there may be associated water blanching of the gastricmucosa). Conversely the absence of liquid in the stomach suggests either rapid death by drowning or death prior to submersion. Debris and chemical contaminants present in liquid recovered from the lungs and stomach can be compared with samples of water from the place of submersion to provide corroboration that drowning occurred at that locale. Microscopy as well as chemical analysis of the gastric contents may be useful in this regard. Vomitus may be found in the esophagus and airways as a result of agonal inhalation or attempted rescuscitation. The presence of large quantities

of sand in the upper airways raises the possibility of inhalation of a thick suspension of sand in seawater produced by heavy surf; death is very rapid in such cases.

Middle ear and mastoid air cell hemorrhage. These are occasionally seen in bodies recovered from water and produce a blue-purple discoloration of the bone of the roof of the mastoid air cells.

Their pathogenesis is unknown and their presence does not contribute to proof of death by drowning. They may be the result of baro-trauma or the irritant/pressure effects of aspiration of fluid into the eustachian tubes or extreme congestion.

Such hemorrhages also occur in cases of head trauma, electrocution and mechanical asphyxiation. Conjunctival hemorrhages Occasional small conjunctival hemorrhages may be seen but the multiple petechial hemorrhages found in other asphyxial deaths are not seen in drowning(except in rare instances of rapid death associated with glottic spasm - see below).The conjunctivae are often congested. Venous congestion and fluid blood - Heart failure combined with blood volume expansion from the absorption of freshwater are reflected in engorgement of the right side of the heart and large veins.

As a result of hemodilution the blood is fluid and thin lacking its normal sticky consistency. Foreign material in the hands. Victims struggling in water may clutch at objects which are then found grasped in the hand after death. Weeds, branches and other objects fixed in the hand by cadaveric spasm (instantaneous rigor) provides good evidence that the victim was alive and conscious at the time of submersion. Similar materials may be recovered from beneath the fingernails.

Injuries to the hands or fingertips and tearing of the fingernails may be produced during attempts to grasp at objects. Shoulder-girdle bruises Victims struggling violently to survive in water bruise or rupture muscles particularly those of the shoulder girdle, neck and chest (most often the scaleni and pectoralis major).

Hemorrhages may be bilateral and tend to follow the lines of the muscle bundles. They may be present in up to 10% of cases and are strong indicators that the victim was alive in the water. In decomposing bodies these hemorrhages should be examined histologically. Uneven putrefaction can cause reddish patches to develop in muscle through hemoglobin inhibition and this maybe confused with hemorrhage. Extra vascular erythrocytes provides histological proof of the existence of true hemorrhage.

ATYPICAL DROWNING

Vagal inhibition (cardiac arrest, laryngeal shock)This is uncommon but well recognized. Loss of consciousness is usually instantaneous and death ensues soon afterwards, at most within a few minutes. Autopsy discloses none of the usual signs of drowning. The mechanism is believed to be cardiac arrest induced by impact of cold water on the back of the pharynx and larynx.

The three circumstances common to these deaths are

(a) entering the water feet first,
(b) surprise or unpreparedness and
(c) a "state of hypersensitivity" e.g. alcohol intoxication.

Entering the water feet first it is easy for liquid to pass up the nose. Alternatively "duck-diving" or any clumsy diving with abdominal impact against the water can produce a similar result. Eyewitnesses observe that there is no struggle by the victim who is found to be dead even if the body is immediately recovered. There may be instantaneous rigor(cadaveric spasm).

Laryngeal spasm
There is likely some element of laryngeal spasm in all drowning deaths. However in these cases there is no evidence of aspiration of liquid and there are the typical signs of an asphyxial death including facial cynanosis and petechial hemorrhages. The mechanism is thought to be sudden chilling of the neck and chest followed by immediate inhalation of water resulting in reflex spasm of the larynx, early unconsciousness and a rapid asphyxia.

The possibility of an asphyxial death prior to entry into the water must be excluded (e.g. homicidal strangulation).

"Dry drowning"
This terminology is not recommended. It derives from the division of drowning cases into "dry" and "wet" according to the condition

of the lungs. However the finding of dry lungs indicates neither that water was inhaled nor that it was not. It is possible that water was inhaled, absorbed into the circulation and then death occurred prior to the onset of active pulmonary edema. The confusing concept of dry drowning has been used to widen the spectrum of cases of vagal inhibition and laryngeal spasm bringing these valid concepts into some disrepute.

Delayed death ("secondary drowning", post-immersion syndrome) Occasionally death occurs after an individual has been taken from the water and appears to have recovered from a near drowning. Autopsy discloses acute pulmonary oedema. This phenomenon has been reproduced in animal experiments. Later complications include pneumonitis, broncho-pneumonia and hyaline membrane disease together with renal failure secondary to hemoglobinuria.

DROWNING TESTS

There are no universally accepted diagnostic laboratory tests for drowning.

Specific gravity of blood - This test was first proposed in 1902. It is suggested that a lower plasma specific gravity in blood from the left side of the heart when contrasted with blood from the right side of the heart reflects hemo dilution produced during the drowning process.

Plasma

Chloride - This test was first proposed by Gettler in 1921. The plasma chloride levels in blood from the left and right sides of the heart are compared. Haemodilution in fresh water drowning is considered to produce a lower chloride level in left heart blood when contrasted with right heart blood. Conversely haemo-concentration and chloride ion absoprtion in salt water drowning is considered to produce the reverse result.

Plasma Magnesium - This test was proposed by Moritz in 1944. High levels of plasma magnesium in left heart blood when contrasted with right heart blood is considered to reflect absorption on that ion from the drowning medium particularly salt water. None of the above tests are considered definitive although some workers believe that they can provide confirmatory evidence of drowning when the body is recovered and the tests performed within a few hours of death. Non-uniform and unpredictable changes in blood electrolytes which always occur after death render these tests less and less useful the longer the interval between death and recovery of the body.

Diatoms
Diatoms or Bacillariophyceae are a class of microscopic unicellular algae of which about 15,000 species are known (approximately half live in fresh water and the other half in sea or brackish water). Classification is based upon the structure of their siliceous valves. The cell structure is unique in that it secretes a hard siliceous outer box-like skeleton called a frustules which is chemically inert and almost indestructible being resistant to strong acids. In 1941 Incze demonstrated that, during drowning, diatoms could enter the systemic circulation via the lungs.

Their presence can be demonstrated in such tissues as liver, brain and bone marrow following acid digestion of the tissue. The use of diatoms as a diagnostic test for drowning is based upon the hypothesis that diatoms will not enter the systemic circulation and be deposited in such organs as the bone marrow unless the circulation is still functioning thus implying that the decedent was alive in the water. The test is limited by the difficulty of excluding the possibility of contamination. Diatoms are ubiquitous in the environment e.g. in the building industry and as dusting powder for rubber gloves.

Additionally diatoms have been found in the organs of decedents not recovered from water, raising the possibility those diatoms may enter the circulation via the gastro-intestinal tract (as contaminants of foods such as salads, watercress and shellfish) or via the respiratory tract (diatoms are normally present in small numbers in the air).The present consensus is that, given adequate precautions to prevent contamination, the demonstration of diatoms in organs such as bone marrow is strong corroborative evidence of death by drowning.

This is true for decomposed bodies provided there is no gross mutilation. It should be confirmed that the species of diatoms found are the same as those present in the water from which the body was recovered (the diatom population varies seasonally with a major peak in the spring and a less pronounced peak in the autumn). A water sample should be obtained with a plankton net (alternatively collect 1-2 liters of water - add a few drops of iodine to kill the micro-organisms, stand overnight, decant with care and retain the concentrate for examination).

Examination of lung fluid for diatoms is of more limited value but their presence in large numbers provides corroboration of death by drowning.

EFFECTS OF IMMERSION

Sinking, putrefaction and refloating

A body in water will usually sink but because the specific gravity of a body is very close to that of water then small variations e.g. air trapped in clothing have a considerable effect on buoyancy. Having sunk to the bottom the body will remain there until putrefactive gas formation decreases the specific gravity of the body and creates sufficient buoyancy to allow it to rise to the surface and float.

Heavy clothing and weights attached to the body may delay but will not usually prevent the body rising. Putrefaction proceeds at a slower rate in water than in air, in seawater than in fresh water and in running water than in stagnant water.

The principal determinant is the temperature of the water so that in deep very coldwater e.g. the North American Great Lakes or the ocean the body may never resurface.

For the Thames, Simpson offers the following guidelines for resurfacing times:
June to August: 2 days;
April, May, September and October: 3-5 days;
November, December: 10-14 days;
January, February; possibly no resurfacing.

At water temperatures persistently below 45°F there may be no appreciable decomposition after several weeks.

In the water the body floats face down with the head lower than the rest of the body so that lividity is most prominent on the head, neck and anterior chest.

Lividity is often blotchy and irregularly distributed reflecting movement of the body in water. It is not intensive and appears a pink or light red color. In cold water it can be dusky and cyanotic. It may be difficult to recognize due to swelling with water of the upper layers of the skin with resultant loss of translucency. Putrefaction begins first within the areas of lividity i.e. the head, neck and anterior chest. It assumes a greenish bronze or dark brown color; if exceptionally dark there is a' tete de negre' appearance. Putrefaction destroys any foam present in the airways and produces instead a reddish brown malodorous fluid containing bubbles of gas which is of no diagnostic significance (a "pseudo foam").

A similar appearing fluid appears in the pleural cavities associated with collapse of the lungs and this also has no diagnostic significance and does not necessarily reflect a pre-existing pleural effusion. Once removed from water putrefactive changes advance with remarkable rapidity. Adipocere which is a soap-like transformation of subcutaneous fat is common in bodies immersed in water usually appearing after some months; it may be present in as little as six weeks. Occasionally bodies found in water develop so-called lime-soap nodules on the intima of veins, in particular the hepatic veins and on the endocardium.

Similar rnodules composed of calcium phosphate, calcium carbonate, neutral fat and proteinacous substances may be found in the skin of bodies recovered from anserina cutis (goose-skin) -This is a roughening or pimpling of the skin resulting from rigor or the erectorpilae muscles most prominently on the thighs. It is of no diagnostic significance since it can occur in circumstances other than drowning, can develop in the interval between somatic and molecular death or can be a post mortem change.

Maceration of the Skin Immersion in water produces progressive maceration of the skin which becomes blanched, swollen and wrinkled. It is first apparent in the skin of the finger pads and then appears on the palms, backs of the fingers and back of the hand in that order. When fully developed it is most striking on the palms and soles. In water 50-60°F early changes can be seen within an hour.

Generally there are obvious changes within 24-48 hours but the process may be delayed for several days in winter. With developing putrefaction the epidermis including the nails peels off like a glove or stocking. Fingerprints may be easily prepared from the glove - reverse fingerprints may be prepared with some difficulty from the exposed dermis. The wrinkling and blanching of water-soaked skin in reflected histologically in water uptake with swelling of the epidermis progressing to epidermal detachment from the dermis.

Tattoos and scars are readily identified in the dermis following autopsy removal of the peeling epidermis. Occasionally chromogenic bacteria (bacillusprodigiosus and bacillus violaceum) invade the dermis of bodies in water after a period of at least one to two weeks and produce patterns giving the impression of tattoos. Post Mortem Injuries Having sunk to the bottom, a body drifting along the water bed will sustain a pattern of injuries reflecting its

head down floating position (see above). Abrasions are typically found over the prominent points of the face, anterior trunk and extremities. A wide range of injuries may be produced by battering against rocks or by passing watercraft in navigable waters e.g. propeller blades (repetitiv eparallel chops).

The body may be attacked by sharks, small fish, sea lice and other fauna. The soft parts of the face are particularly vulnerable to fish and crustaceans.

Injuries may be inadvertently inflicted during the recovery of the body using grappling irons, hooks and ropes. It is particularly difficult to distinguish ante mortem from post mortem injuries because on the one hand water immersion leaches the blood out of antemortem wounds while on the other hand post mortem wounds tend to bleed more readily than usual due to the fluidity of the blood (particularly in areas of dependent lividity e.g. face).

Histological evidence of a polymorphy infiltrate indicates that the injury is antemortem and has preceded death by at least one hour. The absence of a tissue reaction does not exclude the possibility that the wound is antemortem. The injuries must be interpreted in the light of the circumstances buteven so definitive interpretation may be impossible. The presence of pulmonary fat or bone marrow embolism indicates that bony trauma is ante mortem but the absence of fat embolism is not proof that the trauma was post mortem.

The body cools in water about twice as fast as in air (i.e. about 5°F per hour) and reaches the temperature of the water usually within 5 to 6 hours and nearly always within 12 hours.

Courtesy Derrick J Pounder, University of Dundee

References
Davis, J H. Bodies Found in Water, Am. J. Forensic Med. Pathol. 1986 , 7, p. 291 - 297. (good discussion of the investigative approach to these cases).

Spitz, W U. Drowning in Medico-Legal Investigation of Death, Eds. Spitz W U and Fisher R S, C C Thomas, Springfield, Illinois, 1973, p. 351 - 366. (American text presenting, on key issues, widely divergent views to the next reference, a standard British text).

Drowning in the Essentials of Forensic Medicine, 3rd edition, by Polson C J and Gee, D J, Pergamon, Oxford, 1973, p. 440 - 466. (the best and most comprehensive discussion in a British text book).

Copeland A R, An Assessment of Lung Weights in Drowning Cases, Am. J. Forensic Med. Pathol, 1985, 6, 301-304. (good raw data).

Abdallah A M, et al., Serum Strontium Estimation as a Diagnostic Criterion of the Type of Drowning Water, Forensic Science International, 1985, 28, 47 - 52. (the latest in laboratory tests for drowning, yet to be more widely and critically appraised).

Taylor L R, The Restoration and Identification of Water-soaked Documents, Journal of Forensic Sciences, 1986, 31, 1113 - 1118.
(good practical tip, describes freeze-drying method to recover water-soaked identifying documents on a drowning victim).

Diatoms in Taylor's Principles and Practice of Medical Jurisprudence, Ed. Mant A K,
Churchill Livingstone, Edinburgh, 1984, p. 297 - 299.
(good brief review of the diatom test in drowning).

Chapter 13

Hydrodynamics

How Winds Affect Water Currents

Recovery of objects underwater is easy if they are heavy and sink quickly, especially if you know where they went down. What happens if a person drowns and floats off for a distance then sinks? Does it go the same direction? Many believe so and have searched for many hours (and days) while a body was moving (underwater) in the opposite direction. In fact, some bodies have never been recovered, which is devastating for families and loved ones.

Water is like air, it's fluid and dynamic. It does have opposing forces and much like wind, will divert its energy in a direction where there's less friction. But at the waters surface, that's where the similarities end. Take for instance a brisk west wind for 6 hours over a region of lakes that are tied together by a small channel where boats can pass from one lake to another.

As a certified diver, I found it very unusual that on a particularly cold winter day, the winds were blowing from the west and the flow of water (at 15 feet depth) in the channel was almost still. From past dive experiences there have been times where it would be exhaustive just trying to stay in one location without being swept away due to a strong current. Why isn't it happening now?

Well, after being in 15 feet of water for almost an hour (conducting a weapons recovery) the wind finally proved its effects and after pushing the water eastward for an hour, it wasn't long before the flow in the opposite direction occurred, and it was strong. At first I dug my fingers into the soil to keep my position, then starting grabbing large rocks. Finally after losing my grip, I pulled out my large dive knife and dug it into the bottom of the lake and simply held on while a 3 knot flow kept me parallel until I regained enough energy to exit the channel, without being swept into another lake.

Knowing how the water works with opposing wind, over time, can greatly increase your chances of locating a drowning victim or debris field from a shipwreck. But you have to know what to look for in order to achieve success. In researching wind data, you need to go back a couple of days. Simply pulling up wind records for that day won't tell you how the lake was affected that prior day. I say this because after the wind stops, the flow of water continues to move just like the water in your bathtub. Move your hand back and forth and build wave energy. When you stop, that energy continues over a period of time. Lakes are very similar whereas rivers and streams have a continuous flow downstream. Watch and observe eddies where currents can trap a body under a log or along the bank. Oceans are a bit different and you'll need to review the history of hurricanes to see where the strongest push of undercurrents would go. And the Gulf Stream can occupy a book all in itself. A stream of water within an ocean. Objects move very quickly in those types of environment and are more complicated to consider when factoring in wind.

The University of Michigan Hydrodynamics Lab has an excellent computer program to monitor water movement over a long period of time, especially in the open waters of the Great Lakes, oceans too. You can enter in historical weather data to see what direction the surface (and return flow) moves and recreate wave heights.

When you are researching weather/wind data, don't take the peoples comments as factual evidence. You need to research wind data from local airports and farming communities (including hobbyists) who can attest to the wind direction, speed and duration. Print outs of data is critical when doing a micro-analysis of a lake.

The next two chapters deal with wind and water and can help you, the student gain a greater appreciation working an actual case.

Chapter 14

Historical Case Study From The Late 1800's
Re-Writing History on a Ghost Ship

Forensic Meteorology was put to the test as data was limited back in the late 1880's for the Great Lakes. Not like the hundreds of reporting stations available now. Back then it was only 7, with limited and crude data available at best.

Weather Map taken from the US War Department

Knowing how to read the map was also important. Were the winds reporting the "To" direction or "From" direction? Using one map wasn't enough as they were printed twice a day, so we had to go back a few days before the event and print all maps up to the disaster. For those who enjoy watching the History Channel, you may have seen a show called 'Deep Sea Detectives'. This show (Ghost Ship) is where forensic weather altered the anticipated screenplay and became one of the highest rated shows. Less scripting and more ad-lib, which is another way to say it was done 'on the fly'.

History:

The history of the Windiate prior to its last voyage is irrelevant, but for the purpose of this chapter, we'll include it for your reading enjoyment..

Picture 1: Replica of the Cornelia B. Windiate

Thomas Windiate (1827-1911), a native of England, arrived in the new world in 1836 and grew up in Pontiac, Michigan, along with his nineteen brothers and sisters. After his marriage to Cornelia Wallace, he moved to Manitowoc in the mid 1800's and started a career in construction. Although he started in the hotels, his interest changed to harbors
and piers by 1871. Part of his activity also included shipbuilding which ran into 1880. Thomas Windiate eventually merged with Butler Shipyards and this gave him the ability to construct many of the leading boats launched at Manitowoc, one was the *Trumpf*, which had the distinction of being the first vessel to make the trip from Lake Michigan to Europe[1]. Thomas had a strong background in ship construction when he built the vessel "Cornelia B. Windiate". It was this vessel, aptly named after his wife, which is now one of the best preserved shipwrecks in the Great Lakes.

The Windiate traveled from Milwaukee, Wisconsin to Buffalo, N.Y. and back several times, encountering some minor collisions, one of which broke her bow sprit, jib and then her stem[2]. These repairs put the Windiate behind schedule and created a back log of shipping cargo and the result would eventually be catastrophic. Her last voyage of the season, bound for Buffalo, N.Y., she was being heavily loaded with red wheat to make one last run before the late fall weather would settle in, halting shipping until spring.

The lakes harbor heat from the summer's sun and with the cold weather approaching, storms, that would appear docile, would strengthen after passing over the lakes and absorbing the energy released from the heated water. The result is what we now call November gales, hurricane force winds generating waves in excess of 30 feet and confused seas.

However in this instance, it was a different storm, one that would be generated by the extreme temperature difference, creating not only strong gradient of winds but an arctic blast from an extreme high pressure system - Violent weather extremes which set records that still stand today in Massachusetts.

FAIR WEATHER SAILING

With a length of 136 ft. and a width of 26 ft., the "Cornelia B" certainly wasn't built for high seas. With a low draft from a heavy cargo load of 20,000 bu. of wheat, waves would be her downfall despite the A2 rating given to her. The fair skies off to the west indicated perfect sailing into the midnight hours with visibility good enough to keep the Wisconsin shoreline, in view. Loaded heavy with red wheat (this would be a factor in the sinking), the Windiate set sail for Buffalo, N.Y., and with favorable weather to continue and a top speed of 8 knots, the Capt. could maximize his use of winds and make it through the Straits of Mackinac within 36 hours. As the weather map indicates (on a larger full U.S.A. scale), a broad area of high pressure moved from Canada and settled into the northwestern part of the Great Lakes.

What would appear to be nice weather for sailing was about to turn deadly. The Captain didn't know it, but his superior sailing skills were racing the Cornelia B. Windiate into the teeth of a deadly storm induced by the difference from extremely cold air temperatures and warm water temperatures.

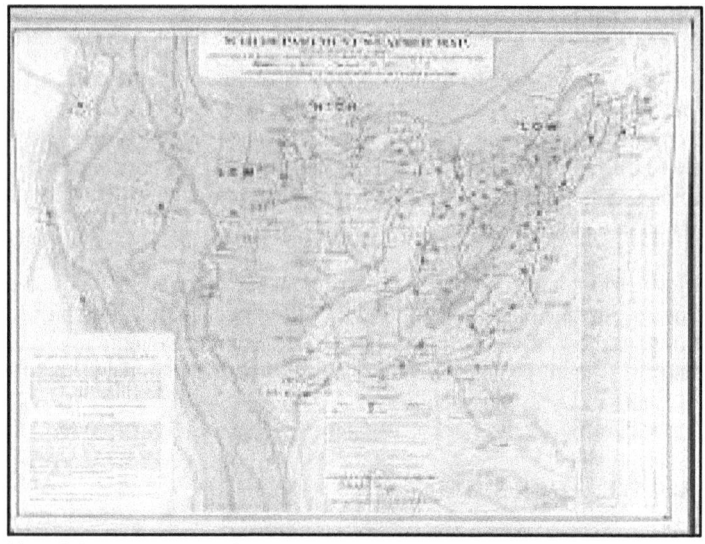

Picture 2: (War Department Weather Map-Nov. 27, 1875)

It was just after noon, November 27th, 1875, when the fairest of sailing weather slowly made its way into the Great Lakes. A Southeast wind at 10 to 15 mph shifting to the West-Northwest gusting to 20 mph. (within 36 hours) provided ample power to keep full sail on a vessel traveling up the shoreline of Wisconsin and Upper Peninsula of Michigan to maintain the best speed possible. It's been recorded in many logbooks that [clear skies and light winds make for perfect sailing] and although the air temperatures were slightly below freezing, the waters showed no signs of icing up.

A RELATIVE TIMELINE MAKES HISTORY

While working with the Deep Sea Detectives in the production of the "Ghost Ship" (Cornelia B. Windiate) – it was apparent, through the use of weather forensics, what had been recorded and copied over the course of 132 years may have been inaccurate. The use of weather recordings were limited in that era and the deep freeze that followed soon after the Windiates' departure would give reason to believe that either it was never seen to pass through the Straits of Mackinac or was caught in the deep freeze, as other ships have and succumbed to the icy cold and foundered to its final resting place, which was assumed to be in Lake Michigan. Newspapers[1] report that other ships were stranded in the ice in the same timeframe of the Windiate, relatively speaking, and shipmates had to wait until the ice thickened enough to walk to safety.

In such conditions, ships hulls would crush under the pressure and sink or, if lucky enough, limp back to the nearest port for repairs. One vessel was reported to have had her sails covered in ice as she settled in for the night. It was the coldest arctic wave to hit the Great Lakes and Northeast U.S. in over 132 years and set records in Boston, MA., that stand to this day. With no record of the Windiate making it to its final destination of Buffalo, N.Y. it was **assumed** to have perished in the deep freeze where its hull would have been crushed or cut up and sunk by the ice. Reports from observers suggest she never made it through the Straits of Mackinac which would indicate why she was assumed to lie at the bottom of Lake Michigan.

1 Chicago Tribune, December 1st, 1985 – Manitowoc Tribune, December 2, 1875

So how did this ship make it into Lake Huron unnoticed?
It's discovery in 1987 by a couple of scuba divers, brought renewed interest in what many claim to be the most well preserved wooden schooner in the Great Lakes. The preservative qualities of the cold fresh water of Lake Huron and the depth of the shipwreck could make time stand still. Despite the recorded documents stating that she had perished in Lake Michigan, the Cornelia B was in fact discovered in Lake Huron. This revelation made the scuba industry stand up and take notice! How did the Windiate defy the newspaper reports claiming she sank in Lake Michigan? A clue was found in the following days of newspaper clippings and was used as a reason why many felt it sank in Lake Michigan, which described in detail an unusual arctic blast of cold air that froze many lakes solid. Hundreds of ships were at the mercy of the ice. Some foundered after their hulls were broken by the ice. This led some to believe that the Windiate met the same fate. After all, it was right around the same timeframe, and if all the other ships met the same demise, why would the Windiate be different?

Greg MacMaster   Environmental Forensics

What follows is a classic example of history repeating itself in books by various authors. Notice the sources referenced, many of which are very dependable.

CORNELIA WINDIATE
Other names   :   none also seen as CORNELIA B. WINDIATE
Official no.   :   2537
Type at loss   :   schooner, wood, 3-mast
Build info   :   1873, Thos. Windiate, Manitowoc, WI
Specs   :   136x26x12, 322 t.
Date of loss   :   1875, Dec 10
Place of loss   :   off Middle Isl. near Rogers City, MI
Lake   :   Huron
Type of loss   :   storm
Loss of life   :   9 [all]
Carrying   :   wheat
Detail   :   Bound Milwaukee for Buffalo, she became trapped in ice, cut and sunk. She was not reported as having passed the Straits and her spars were reported sticking out of the water near the Fox Islands, so she was thought for over 100 year to have been lost in Lake Michigan. She was discovered on the bottom of Lake Huron in 1987, in excellent condition.
Sources   :   diver,glss,usls,mpl,es,nsp

The source list was referenced from: http://greatlakeshistory.homestead.com/files/sources.htm and listed as: diver, **glss** - *Great Lakes Shipwrecks and Survivals* - Ratigan, William - Eerdmann's Publishing, Grand Rapids, MI, 1960 (still in print: Freshwater Press, Cleveland) - **usls** - United States Lifesaving Service annual reports, **mpl** - Milwaukee Public Library Great Lakes Marine Collection (including Runge file summaries) - Web page at:

If you did a Google search on the Cornelia B. Windiate, you would find various reports with two fundamental historic conclusions:

1. It sank in Lake Michigan because it wasn't seen to pass through the Straits of Mackinaw and;
2. It was chronologically associated with extreme cold weather, thus similar to other vessels at the time, the Windiate sank after it became stuck in ice and hull ruptured.

The strange thing was that modern divers did not find any scratches or any holes in the hull as was the case in the Deep Sea Detectives where John Chatterton and Richie Kohler reviewed their dive video with the author.

UNIQUE OUTLOOK

Let's examine some of the potential reasons that the location of the Cornelia B was assumed to be in Lake Michigan. Two weather events could have easily contributed to the lack of a visual sighting of the Windate passing through the Straits at midnight. The first was the presence of lake effect snow. Little was known at that time about the phenomena of lake effect snow, a phenomenon that develops when very cold air rides over warmer waters. If conditions are right and there is a significant temperature difference between the air and water, heavy snow can develop along the shoreline of Michigan under a light westerly wind, rendering visibility down to zero at times. Lake effect snow is fairly uncommon in the rest of world but a normal weather event in the Great Lakes every winter.

This phenomenon makes it easy for vessels to travel outside of the cloudbank in perfect visibility and cloudless skies, yet shoreline observers on the leeward side couldn't see any distance at all. If the Windiate made best speed and stayed outside of the cloud bank that harbored heavy lake effect snow over the shoreline of Michigan, it could have easily sailed right through the Straits of Mackinac unnoticed towards the midnight hour of November 29, 1875.

Another critical weather event that was occurring at that time was the immediate drop in temperature which was falling from the low 40's to the single digits within 6 hours. This dramatic change in temperature created very strong winds and rough seas. Could it be that it was assumed that no ships would attempt to pass through this rough weather? This seems to be the conclusion of the regional newspapers at that time. Combined with a lack of visual sightings people truly believed sank in Lake Michigan. This erroneous conclusion grew to be the recorded word, used by authors, historians and archaeologists to this day. The Windiate was sailing parallel to the Upper Peninsula shoreline which slowed the strong winds due to friction.

The problem with previous history attempts, the evidence did not fit this scenario as the vessel did not have a scratch on her, no damage to her top deck, cloth hatches and pilot house not blown off and anchor chain laid out neatly? Almost every ship that sinks would break in half, land upside down, have parts of the deck blown off due to excessive air pressure, some kind of damage. Yet, to have the chain out as if it was waiting for a storm raised questions that needed further investigation.

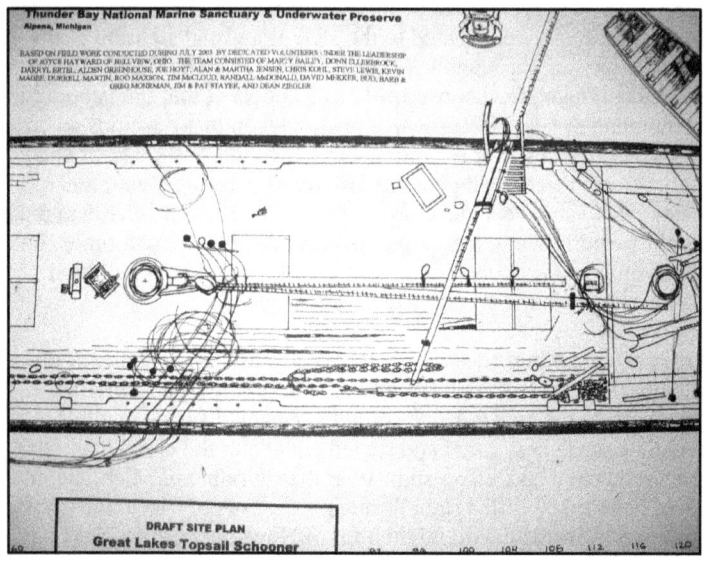

Picture 3: Draft Site Plan, Great Lakes Topsail Schooner, Windiate.

Courtesy: C Patrick Labadie

Aside from the production efforts of the Deep Sea Detectives, what transpired in a conference room with C. Patrick Labadie, Historian and Greg MacMaster, Forensic Meteorologist and author, ended up not only changing the production script, but re-writing history on the fly. Earlier in the day, too many unanswered questions brought even more questions. If it did encounter ice, where were the holes in the hull? Were there and gouges along the waterline to indicate ice damage?

Picture 4: Pat Labadie & Author, Greg MacMaster

Also, why was it laying on the lake bottom perfectly intact?

As Patrick and Greg plotted the route of the vessel, the time it left port and the re-creation of weather events, they were able to plot an estimated course and determine that, if the captain was aggressive enough, he could have sailed right through the Straits.

Pat Labadie agreed that if capt. maintained a good course, he could have easily traveled at 8-10 knots. This would bring the Windiate through the Straits of Mackinac right around midnight on November 29, 1875. Everything looked good, other than a drastic drop in air temperature. It appeared the captain utilized the winds, which were picking up from the northwest, gusting to 35 knots (yet shielded from the landmass to the north) and to his advantage he was ready to cruise into Lake Huron. Temperatures cool into the overnight hours, so the Captain most likely shrugged it off. That drastic temperature difference between air and water was creating a rage in winds and waves in Lake Huron, waiting for the Windiate.

SCIENTIFIC EVIDENCE

Weather data for November, 1875 was scarce with only a handful of observations in the area of interest, taken two or three times a day. Not like today's technology where an observation can be taken every 5 seconds. Information in the late 1800's was crude in coverage, but could yield results through a microanalysis (a more concentrated area in relationship to the larger picture taking into account all of the weather data) of weather elements. What we needed was wind speed and direction, temperature and pressure for locations around the Great Lakes, specifically the port of departure, northern Lake Michigan, Straits of Mackinac area and Alpena region. Unfortunately, the stations we had to work with were slightly off the mark I was hoping for. However, having land-based weather observation stations in Wisconsin, Michigan, Canada, Ohio and Illinois filled in the gaps to make a more complete analysis and formulate a conclusion.

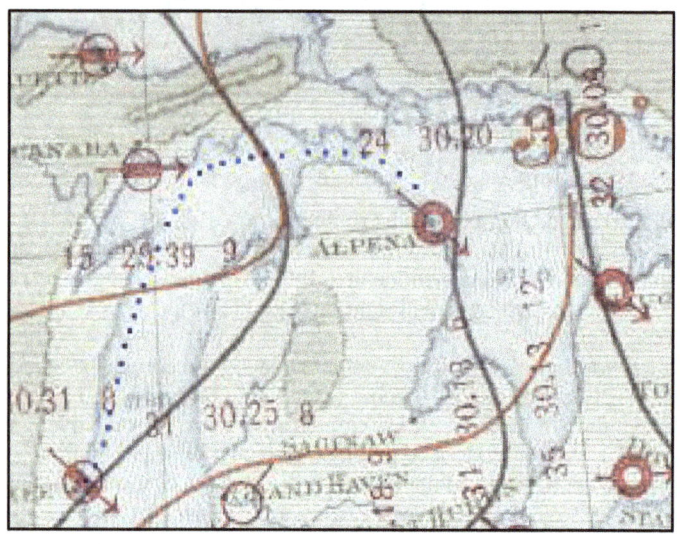

Picture 5: War Department Weather Map-Nov. 27, 1875 – Close up view showing area of interest (blue dots show path of vessel)

The weather map (picture 5) shows the direction of the wind flow, temperature in F. and pressure in inches of mercury. Using 3 maps in a 24 hour period it was time to make a time-line to show a cross section of the path traveled. Once Pat Labadie penciled in his best guess based on the weather conditions (blue dots), Greg was able to create a cross section of the Windiate path in comparison to the time of day and added weather reports. Here's what we discovered through the use of forensics;

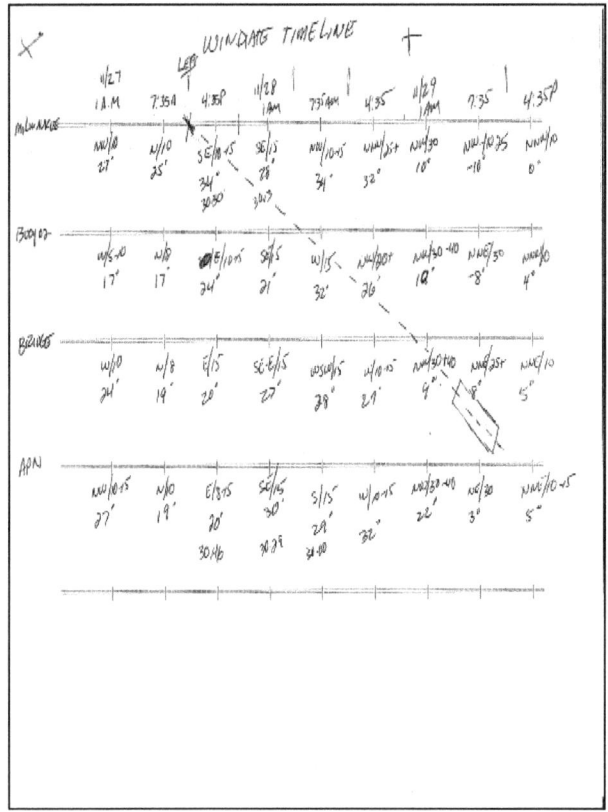

Picture 6: Draft timeline of weather events from Milwaukee to wreck position. Source: Greg MacMaster

1. The Windiate left port in Wisconsin under ideal SE winds and temperatures close to freezing.

2. The winds shifted to the west-Northwest as the Windiate rounded Summer Island, just south of Fairport, MI in the Upper Peninsula.
3. Wind speeds increased to 25 mph from the Northwest, yet the Windiate was somewhat protected compared to her position south of the Upper Peninsula, thereby shielding her from increasing waves – yet still maximize speed.
4. As she passed through the Straits (Air temperature of 32°F), with stormy weather, many would probably be seeking shelter as it was close to midnight.
5. With a quarter wind on her back port side, the Windiate rushed through the Straits and shielded from the landmass to the north and passing Bois Blanc Island (Air temperature 15°F), she headed into the beginning stages of the coldest arctic blast of winter weather that would stand for 132 years. The extreme cold and wind was generating significant waves.

6. Considering the anchor chains position on the deck (See artist graphic), the Windiate was readying herself for rough seas.

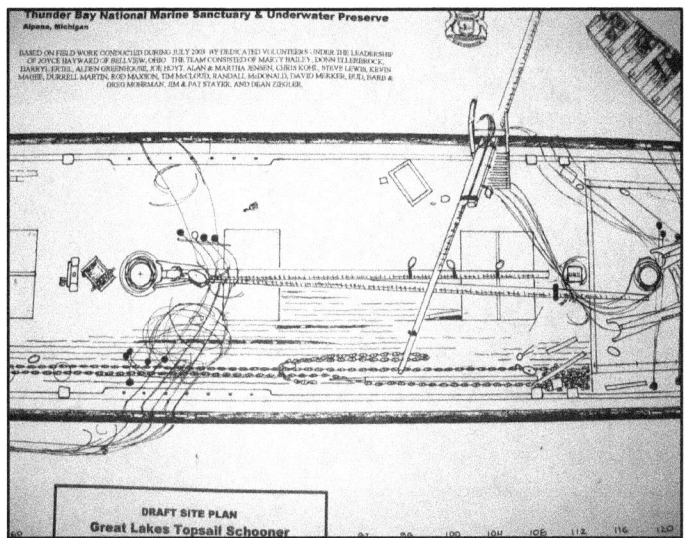

Picture 7; Draft Site Plan, Great Lakes Topsail Schooner, Windiate.
Courtesy: C Patrick Labadie

THE FINAL HOURS OF THE WINDIATE
Depending on the wind direction and speed, a land mass can be an excellent protector against large waves, however. Once you leave the protection of the land, you become exposed to the full force of the elements, which can be just as treacherous as the open seas. As the Windiate passed the Straits, the temperature was cooling to 32F, by the time it passed the eastern
point of Bois Blanc Island, the winds shifted to the northeast and increased to 30 mph. With her anchor line ready –

Picture 8: DSD rendition of Windiate

she continued on as cautiously as she could. Little did the captain know that running along the shoreline from Rogers City to Presque Isle Light House only increased the fetch of the wind, thereby increasing the waves. Air temperatures were down to the single digits and as winds increased, wind driven spray was coating the vessel from the waterline up. Remember, the Windiate left Milwaukee in temperatures below freezing, so every part of the exposed vessel was becoming super cooled. Once the wind driven spray touched the vessel, it froze on contact. In essence, the vessel was increasingly coated with ice while it sailed down Lake Huron.

This coating of ice adds weight to the already heavily loaded vessel and it slowly loses its free board (the distance from the waterline to the deck), and forces the ship down due to gravity. Combined with waves of 15-25 feet, and even higher in extreme temperature differences and you have a vessel caked in ice floundering helplessly. If you can picture standing on 3 inches of ice, going to your car with an ax and with gusting winds of 30 mph and freezing rain – try breaking ice away. Now add the unbalanced deck of a vessel in those conditions and you'll get a pretty good idea that these conditions would seal their fate.

Picture 9: DSD rendition of Windiate Sinking

Regardless of whether the crew was swept overboard or went down with the ship, what we do know is that the ice covered Windiate was slowly sinking. That means everything on the deck was frozen into position and would remain there to this day.

This includes the covered hatches made of cloth that would have blown out otherwise; the pilot house which would have broke apart from the escaping air from below deck and the box, chains and other deck items. The ice was heavy enough to push her below the surface yet provided enough buoyancy to slowly glide her down to the bottom of Lake Huron virtually unscathed.

To help reduce errors in the conclusion, John Chatterton and Richie Kohler were inspecting the hull and working with the University of Michigan Hydrodynamics Lab on the ROV. They also inspected the stern rail for evidence of ax cuts, keeping in mind that those on board probably weren't expecting to encounter such harsh conditions – otherwise, they would have laid up and waited out the storm. There were no ax cuts and no visible damage to the hull – the Windiate never had a chance. The position of the yawl boat, still close to the vessel, suggested the crew were not even aware of the danger or conditions were so bad that they couldn't go top side for fear of being blown overboard.

Picture 10: Yawl Boat rests next to the final location of the Windiate

Source: Bob Thorpe

After diving the wreck, John Chatterton, Richie Kohler and Greg reviewed what we saw on video and agreed that there was no way this vessel could have succumbed to ice damage from being stuck. For one thing, the weather prior to this arctic outbreak was typical for fall weather. The temperatures earlier in the summer/fall did not support the theory that the Windiate had become stuck in ice and had its hull crushed. The water was too warm. There were just too many inconsistencies with historical reports to suggest that scenario

CONCLUSIVE EVIDENCE SUPPORTED BY FORENSIC ENGINEERING

At the University of Michigan Hydrodynamics Lab in Ann Arbor, Michigan, Guy Meadows (from the Marine Hydrodynamic Laboratories, University of Michigan Department of Naval Architecture and Marine Engineering) and Hans VanSumeren (who now is the Director of Northwestern Michigan College Water Studies Institute, stated the following in an e-mail; "We ran a fetch limited wave growth model originally developed for the Great Lakes by Mark Donelan, formally of Canada's Center for Inland Waters and then later modified by the folks at the NOAA, Great Lakes Lab, here in Ann Arbor. We used the weather reports provided (by the forensic meteorologist) and used our standard correction for over water winds.

This brought us up to the 50 mph range reported. The large waves also resulted from the extremely strong Air-Sea temperature difference. When this value reached the -30 range (Air temperature − Sea temperature), waves grow very fast.

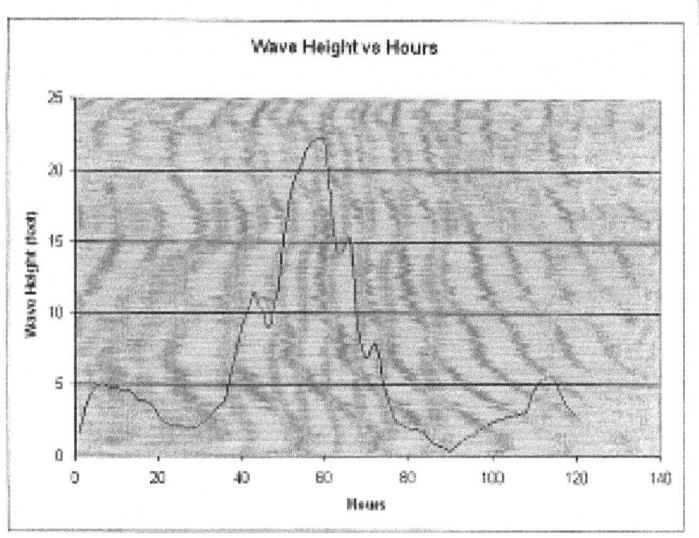

Picture 11: Wave Simulation program Output: Source U of M.

These conditions produce waves approximately twice as big as would be expected under normal conditions similar to those that sunk the Edmond Fitzgerald. They also produce waves that grow very fast. Hence, we believe the intense outbreak of very cold weather contributed to the wreck in two ways; i) rapid and intense wave growth and ii) conditions very conducive to ice formation above the waterline."

Picture 12; University of Michigan Hydrodynamics Lab Wave Simulation

To test this hypothesis a vessel weighted to simulate the accumulation of ice and free board was placed in a wave pool and the weather data was applied. In the following pictures, you'll see the Windiate *replica* in a series of time lapse photographs in the wave pool where she eventually foundered and, as expected, tipped forward and descended to her final resting place.

Picture 13
Simulation of 25 ft. waves

Picture 14
Simulation of
30-40 ft. waves

Picture 15
Windiate sinks

SUMMARY AND
CONCLUSION:

The layout of the chain on the Windiate (and other items) and lack of damage to its hull from ice indicate that previous attempts to portray its demise were wrong. Using science, specifically **forensic meteorology**, to back-track the weather and compare it to other newspaper reports, gave credence to the theory that the Windiate did in fact travel through the Straits undetected and encountered significant waves from sustained storm force winds in northern Lake Huron. The hyper cooling also suggested that considerable ice accumulated on the deck and other exposed areas (a gallon of ice weights approx. 6 lbs) offsetting the buoyancy of the ship and cargo. The cargo of red wheat proved to be a factor in the slow sinking because red wheat, when water-logged will actually float for a period of time settling to the bottom. After a sample extraction of red wheat from the hull, it was sent out for a DNA test for verification.

With a length of 136 feet and a width of 26 feet, the estimated surface area of decking would be 3500 square feet. The estimated weight of ice would be in excess of 50,000 lbs just on the deck, not including the masts, rails or anything above the deck. With the top deck sealed in ice, the sinking was slowed and the Windiate slowly descended to its final resting place virtually untouched by the storms fury. Once the ice melted away, everything on the Windiate was "frozen in time" (see Figure 1) giving the nickname "Ghost Ship".

Without the use of forensic weather, the history books would have continued with the traditional popular stories published to this day.

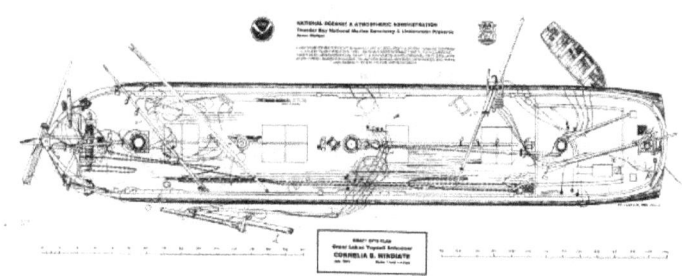

Figure 1. Cornelia B. Windiate (Artist rendition courtesy: C. Patrick Labadie)

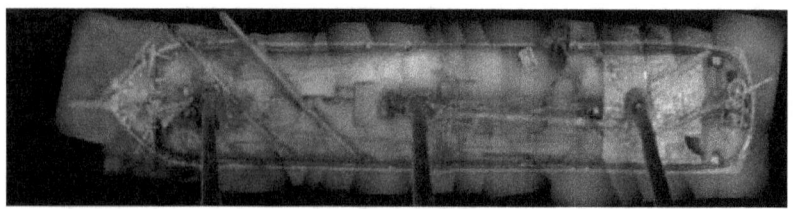

Figure 2. Mosaic work completed by Thunder Bay National Marine Sanctuary and NOAA

Greg MacMaster Environmental Forensics

Wisconsin Maritime Museum, Anchor News (Winter 2006)

Chicago Inter-Ocean Newspaper, courtesy Cris Kohl Collection

Chicago Tribune, December 1st, 1985 – Manitowoc Tribune, December 2, 1875
In reviewing different books on the history of the Windiate sinking, many sources were referenced: The source list was referenced from: http://greatlakeshistory.homestead.com/files/sources.htm and listed as: diver, glss - Great Lakes Shipwrecks and Survivals - Ratigan, William - Eerdmann's Publishing, Grand Rapids, MI, 1960 (still in print: Freshwater Press, Cleveland) - usls - United States Lifesaving Service annual reports, mpl - Milwaukee Public Library Great Lakes Marine Collection (including Runge file summaries) - Web page at: http://www.mpl.org/files/central/greatlakes/glakes.htm - es - Echo Soundings (Amherstburg Echo) Vol. 1 Nos. 1-4, Vol. 2, No. 1-3 - Marsh Collection Society, Amherstburg, Ont., 1998-99.
Picture 1: Replica of the Cornelia B. Windiate – Photo Courtesy Greg MacMaster
Picture 2: (War Department Weather Map-Nov. 27, 1875) - Courtesy NOAA
Picture 3: Draft Site Plan, Great Lakes Topsail Schooner, Windiate. Courtesy: C Patrick Labadie
Picture 4: Pat Labadie & Greg MacMaster – Courtesy Deep Sea Detectives
Picture 5: War Department Weather Map-Nov. 27, 1875 – Courtesy NOAA
Picture 6: Timeline of weather events from Milwaukee to wreck position. Source: Greg MacMaster, Forensic Meteorologist
Picture 7; Draft Site Plan, Great Lakes Topsail Schooner, Windiate. Courtesy: C Patrick Labadie

Picture 8: DSD rendition of Windiate - Courtesy Deep Sea Detectives

Picture 9: DSD rendition of Windiate Sinking - Courtesy Deep Sea Detectives

Chapter 15

Search & Recovery
Mullet Lake Drowning

A drowning took place on a beautiful lake in northern Michigan. Mullet Lake situated in Cheboygan County. It's a fisherman's paradise and an excellent place to have a summer cottage. With a small population, everyone knows everyone and this small town appeal was shattered one night. It was dinner time on June 10th, 2009 when two adults in their mid 30's headed out for a canoe ride along the western side of Mullet Lake

I had scheduled a meeting with the Sheriff of Cheboygan County on an unrelated subject (which was cancelled that day due to the forgoing investigation) and it was by chance that I happened to make the drive to Cheboygan anyway only to see a dive team I often trained with, out on Mullet Lake in a search pattern. I pulled in and asked to speak with the Sheriff and after a brief introduction; we discussed the case in detail. The Sheriff explained that the debris field was quite extensive and larger than what he is used to seeing on drowning cases. I agreed and we quickly discussed alternate investigations in financial conditions to determine if the alleged victims may have fled instead of drowning.

I asked the Sheriff for the latitude and longitude of the debris field and I was willing to offer my forensic expertise in where the bodies would be if they did in fact drowned. Considering my other employment, I assured him that I would keep any information confidential. Keep in mind that the search team had already been on the water searching from the previous night. I didn't start my wind analysis until Friday (June 12th, 2009)

In map 1, you'll see on the east shoreline a blue vest, orange vest and canoe. On the west shoreline you'll see an orange hat and blue coat and the last known location of the 2 people.

The investigation team headed out to search for the bodies on June 11, 2009 and interviewed neighbors and called out the U.S Coast Guard for a preliminary fly over the lake to look for any signs of survivors. It was then that they discovered the following:

Greg MacMaster　　　　　　　　　Environmental Forensics

Item	Found at:	Location:
Blue coat	11/1950 hrs	N 45 28.555
		W 84 35.611
Blue life vest	11/2045 hrs	N 45 29.18
		W 84 31.65
Orange life vest	11/2045 hrs	N 45 28.93
		W 84 31.87
Orange hat	11/2048 hrs	N 45 28.329
		W 84 35.634
Swamped canoe Found on the 11th	p.m. sometime	N 45 28.228
		W 84 33.098

These debris items were found on (June 11th, 2009) the day after the couple went missing.

If you are using this chapter as an exercise, go to www.wunderground.com and type in the zip code 49721.

Search in the history section for June 10th, 2009 and the following day. Print out the two days and start a wind analysis over the lake using a streamline analysis. If you are unsure how to start a streamline analysis, get with your instructor or meteorologist for a refresher. You now have all the necessary data to make an analysis of wind flow prior to and hours proceeding the drowning.

Many investigators would ask what the winds were for each day not realizing that the fluid dynamics of the wind would interact with water and create an opposing flow (undercurrent) which would take the bodies in the opposite directions. Here's your chance as a forensic weather professional to help the recovery team locate and retrieve the bodies and bring closure to the family.

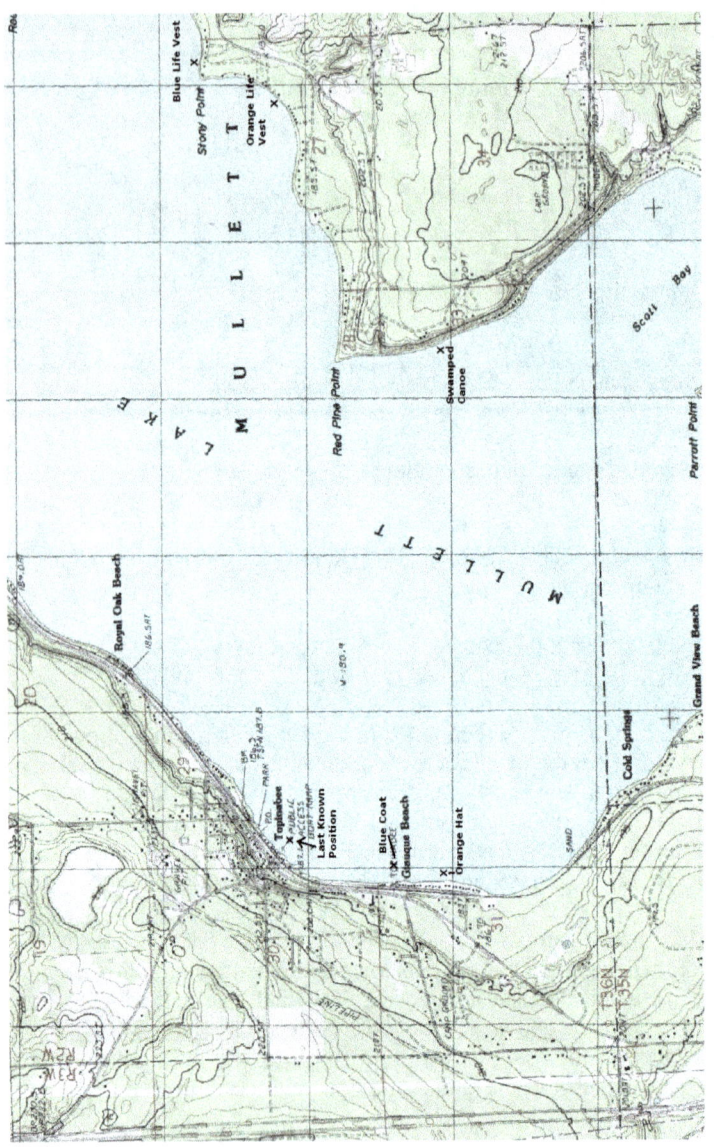

Map 1: Mullet Lake with debris items marked by an 'X' and originating location.

If you are using this chapter as an exercise with students, take the map and make copies (or re-create from Google Maps or other terrain maps) and gather the necessary weather data from the previous source listed and start the analysis.

If you are casually reading this, here's the theory that put it all together.

A strong east wind 10 to 25 mph on Tuesday over the lake created a fairly strong undercurrent in the opposing direction. Tuesday night, the winds die down however the undercurrent is still present. The timing of the canoe tipping over (by the yelling of the people) looks like they headed out as winds were slowing down.

On Wednesday the winds shifted to the west with gusts to 20 mph by mid day then shifted to the West-northwest at 2:15pm. You now have an undercurrent and wind flow going in the same direction.

If you took a floating object (like a hat and jacket) and immediately threw it in the water, the east winds would carry that article of clothing to the west side of the lake where it would be fairly protected by trees if an opposing wind would blow (which it does later). If you fell off a canoe, unless you are knocked unconscious, you would grab a life preserver and stay by the canoe for as long as possible. Due to the water temperatures being in the 50's, it wouldn't take long before hypothermia started to set in. Knowing how the canoe was found (swamped) leans to the belief that they tried to get back in and failed thus swamping the canoe.

Nightfall sets in and air temps coupled with water temperature certainly accelerated their fate. With 95% of the body submerged in water, the undercurrent will have more control over the direction of the bodies (at first). As Wednesday's strong winds approached, the floating objects would continue on a eastward path until they landed on shore (see the blue and orange life vest), which was virtually due east from the last known location where they tipped over. This would make sense with strong west winds pushing them across the lake. However, look at the location of the canoe. Its farther south than the other items found on the east side of the lake. Considering it was found swamped, the travel time would have been slower due to the frictional effects of the water and made it more susceptible to a WNW wind flow which is why it was pushed father down the lakeshore.

So back to the hat and jacket on the west side of the lake; With a strong wind over the trees, often times you'll see a swirling flow creating a mini vacuum keeping items closer to shore since they are not fully exposed to the strong west winds. Now we have the reasoning as to why the items are where they are, but how do we tie them all together?

If you draw a line from each debris item found back to the starting point (where they tipped over) and arc the canoe line to reflect the west-northwesterly component of the wind, you'll start to see a pattern of where to start looking. Once a body is submerged in water (unless its in a river or stream) the speed it moves is very slow.

So let's put this into perspective: Both adults fell off the canoe and their hat and jacket floated away with the surface water movement. They clung to the canoe and grabbed life preservers in hope to stay alive and tried unsuccessfully to get back into the canoe and started floating off to the east with the undercurrent. The person who expired first probably had the blue life vest as it was exposed to the west wind component the longest and the person who had the orange vest held on longer and thus was less exposed to the west-northwest wind component. This is merely speculation as to why objects landed where they did and why they're grouping was so scattered. As they drifted away from the canoe, they sank to the bottom and it is there where the undercurrent takes over.

If you look at Map 2. You'll see a red box drawn where I believe the bodies would be headed. Consider the bathymetry of the bottom of the lake. If the body is no longer suspended in equilibrium, it will sink and follow the lower part of the lake which happens to be the same direction as the undercurrent.

The search team has been out for a day and a half looking in the same general location where the hat and coat was found and after I submitted my best guess and followed up with a phone call with "take your starting point and head towards the orange life vest, drop the tow-fish in 12 feet of water and you should run right into them heading towards deeper water". The Cheboygan County Sheriff called to say they had located the first body with 45 minutes and sent divers down to retrieve the body. I mentioned that the same path should be maintained and the second body should be found fairly soon and within the next hour, the second body was located.

While gratifying as it is to be helpful in this type of investigation, I wanted to take something from this case and help pass along the success to others in hopes to speed up recovery investigations.

Knowing the elements you are working in certainly helps in an investigation and being a certified diver (and practicing underwater archaeologist) working closely with law enforcement on exercises did help sharpen my ability to recognize elements that are often overlooked by other meteorologists. If you are unsure about another science, like hydrodynamics, aviation or agriculture – get with local professionals who are experts in their field and learn as much as you can relying on their input. Over time you'll gain a greater appreciation for their profession and the true art of forensic meteorology.

Formulas to assist n finding submerged bodies

Step 1

> Depth = (D) in feet
> Estimate rate of descent for a human body = 1.5 ft/sec.
> D ft. / 1.5 ft./sec. = X seconds

Step 2

> Speed of current = (V) in ft./sec
>
> (X)*(V) = distance body will travel from point last seen

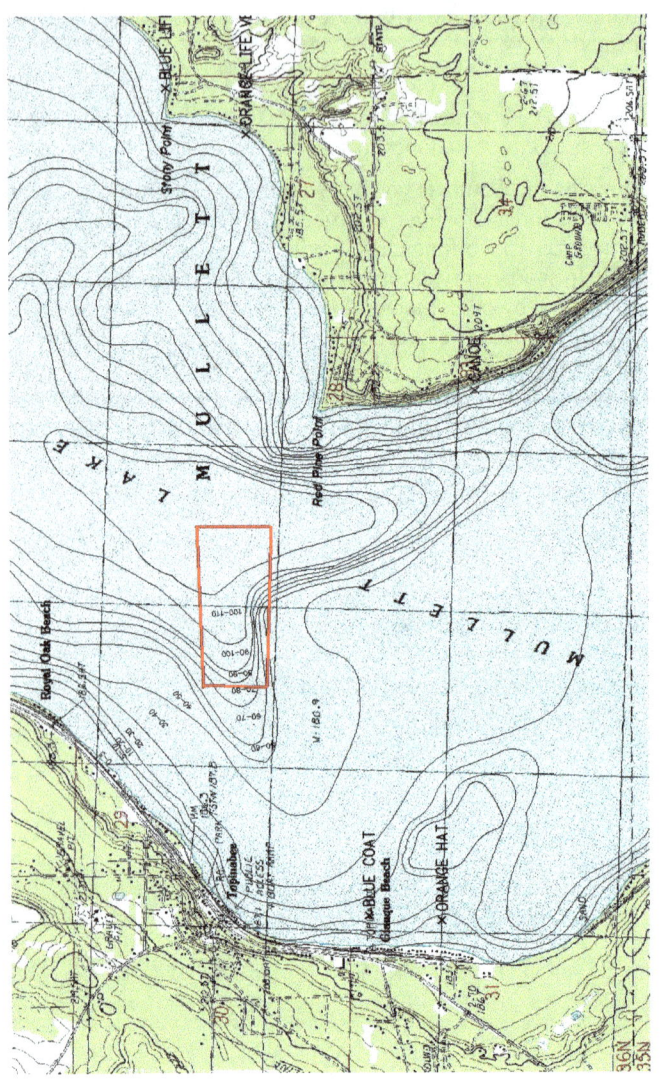

Map 2. Mullet Lake with search box

Chapter 16
Sample cases from the field of forensic & meteorology

Some of the cases below are from various sources and does not mean that the author endorses them. We appreciate all material that has been submitted and they have all been credited where appropriate.

The cases below are real-life examples of how a forensic meteorologist can help make a difference in the legal realm.

A 24-karat weather verification

A jewelry store in the southwestern part of the United States ran a contest that stipulated if the temperature reached a certain record level on the day of the contest, that every customer who purchased from the store that day would receive free jewelry. The record temperature on the day of the contest was 115 degrees. According to the jewelry store, the temperature that afternoon hit 116 degrees, which meant that every single customer was given free jewelry, costing the store some $100,000!!

Prior to the contest, the store had taken out an insurance policy to cover their potential loss, should the record-breaking event occur. And according to the store, it did. They even turned in an "official looking" document to their insurance company, claiming they had a loss which they wanted to be reimbursed for. The claims adjuster for the insurance company contacted a forensic meteorologist and requested a verification of the temperature for that day. The report provided to the insurance company by the forensic meteorologist showed a temperature of only 112 degrees. The reporting station that the jewelry store used in their claim to the insurance company was from a location which is not monitored or controlled by The National Oceanic and Atmospheric Administration. Since there is no quality control at this site, it is not considered an official site for weather records. Based on the report, the claim was denied saving the insurance company $100,000.

Was it an event-in-progress?

Plaintiff, a woman in her 30s, filed suit against a hospital in southeastern New York, after slipping and falling due to a puddle of water at the entranceway to the building. The woman was entering the building with her boyfriend, who was there undergoing blood tests. Her boyfriend testified that he noticed the puddle at 9:20 am as they were entering the building. They both claimed that the accident occurred as they were leaving, and that it was not raining or snowing at the time of the slip.

Defendant Hospital contended that the accident occurred as plaintiff was entering the building, and that a snowstorm was still in progress at the time. At trial, expert testimony from a forensic meteorologist, confirmed the Hospital's stance that snow was indeed still falling at 9:20 am, and was part of a storm which had started several hours earlier. The storm came to an end shortly before 10:00 am, which was determined to be after the time that the slip took place.

Verdict came back in favor of defense, based on the forensic meteorologists' testimony that at the time of the slip, snow was falling, and thus, it was an event in progress.

When did the hole become visible?

Plaintiff, a woman in her 40s, filed suit against a horse farm located in the Catskill Mountains of New York. She sustained severe injuries after the horse she was riding on, stepped into a hole located on a riding trail inside the expansive farm, and fell over, causing her to be thrown to the ground. The trail is maintained by the farm, and is open year-round. The incident took place in the month of December.

Defendant Horse Farm contended that they had no knowledge of the hole's existence, as it was filled and covered by snow and ice from winter storm events which had occurred in the two week period leading up to the day of the incident.

At trial, the forensic meteorologist testified that it had indeed snowed on two different occasions in the 10 day period leading up to the day of occurrence. However, temperatures and degree of sunshine during subsequent days after the last snow event would have been sufficient to cause the amount of snow that had accumulated, to melt away completely. And that this would have taken place at least 60 hours prior to the date and time of the incident.

Verdict came back in favor of plaintiff, based on the testimony that more than two days time would have passed in between the time that the hole became visible after the snow melted away, and the time that the accident occurred.

Real-life examples provided by Compu-Weather.

The cockleburs on a ski-mask, that convicted a rapist

One midnight in midsummer in a suburb of Chicago a woman parked her car and walked toward her apartment building. Suddenly a man wearing a ski mask leaped from the shrubbery, attacked her, and then disappeared. The police began to suspect one man in the building, and with a warrant they searched his apartment and found a ski mask, which he claimed he had not used since the previous winter. The victim identified the man in a voice lineup, but this was not enough for a conviction. There were two cockleburs stuck to the ski mask and the detectives sent them to forensic entomologist Bernhard Greenberg for examination. Within the cockleburs were live weevil larvae. Examination of the cockleburs found on the crime scene proved to be of the same species as the weevil found on the ski mask. The species was identified as *Rhodobaenus 13-punctatus* Illiger, and is also known as the billbug. This species has a 1-year life cycle, and the larvae pupate in the cocklebur and emerge in the latter part of the summer, and then hibernate. Larva do not over winter, and they would not survive the winter within a desiccated cocklebur in a heated apartment. The suspect was then caught in a lie. The court trusted this evidence, and the rapist was convicted.

The chigger bites that convicted a man for murder

In 1982, deputies of the Ventura County Sheriff office noticed that a murder suspect had chigger bites similar to the ones investigators at the crime scene had on their waste-lines, ankles and behind the knees. The entomologist Jim Webb was contacted, and by analyzing the bites, Webb connected the suspect to the crime scene where the naked body of a 24-year-old woman was found on August 5, 1982. She had been strangled with her own blouse.

They did several tests at different places, but the only place they found chiggers in was a narrow strip near a eucalyptus tree under which the woman had been found. This meant that the suspect had to be at the crime scene at some point, which did not correlate with his testimony. The suspect claimed to have seen the woman the last time at a bar. The suspect was convicted for first-degree murder and sentenced to life without parole.

The partly submerged woman in the sandpit

The dead body of a woman was found in an old sandpit in the city of Helsinki, Finland on 27 June 1964. The sandpit was filled with water and the body was partly submerged and covered by bits of board. Presumably, it had earlier been completely under water. It was largely decayed, but still retaining its original shape. The soft parts of the face had decayed away. Adipocere formation had occurred. At autopsy on 29 June 1964, some fly larvae and puparia were detected in one hand. The larvae were dead and dry by the time they reached the Zoological Museum for examination. In spite of this it was possible to see that three of the larvae belonged to the genus *Muscina* and one to the genus *Fannia*. No flies or fly parasites had emerged from the puparia. No blowfly larvae or water insects were found.

Conclusions based on entomological findings:

The occurrence of larvae and unhatched puparia in the body indicate that it had been accessible to flies for at least one week, because in the prevailing cold micro-climatological conditions the development of *Muscina* fly egg to puparium takes about that time or a bit longer. The absence of blowfly larvae, which attack bodies in their initial phase of decay, indicated that the body had not been accessible to blowflies at that stage. Meanwhile, decay had advanced so far that the corpse was no longer attractive to blowflies, but was attractive to *Fannia* and *Muscina*. Hence it was concluded that the body had been completely submerged for a comparatively long time.

Validity of conclusions:
Police investigations showed that the woman in question had been murdered in the middle of July 1963 and had been hidden by the murderer in the sand-pit. Therefore the conclusion based on the absence of blowflies, although valid, had only a low degree of accuracy. The conclusion based on the occurrence of larvae or puparia of the genera *Fannia* and *Muscina* was indicative of the time during which parts of the dead body had been accessible to fly oviposition, but was of no significance for the determination of the time of death. Obviously, the hand had first emerged from the water and the flies had then oviposited on it.

The body in the bed

The body of a woman was found in her bed in a flat in the centre of Helsinki, Finland on 1 September 1965. Death had obviously occurred on 10 August, since newspapers had not been removed since that day. This was a case of suicide with sedatives. The cadaver was moderately decayed and greenish. The skin surface was loose and the viscera decayed. Blowfly larvae emerged from the orifices of the body. Autopsy was performed after regiration for one day and the fly larvae were collected on this occasion. The larvae were in bad condition at the start of the rearing, except for two small specimens. Rearing at room temperature yielded two small imagines of *Calliphora vicina* on 27 September.

Conclusions based on the entomological findings:

The woman in question had been dead in more than 7-8 days, because fly oviposition occurs on about the second day after death, and development to mature migrating larvae takes about 5-6 days.

The cadaver had been in shadow, because the scotophilic *C. vicina* had oviposited on it, but *Lucilia* species had not, although they were still active.

Validity of conclusions: Both entomological conclusions were validated by the known facts. The conclusion that more than 7-8 days had elapsed since death is true as such, but of little value, because death had in fact occurred 20 days before discovery of the cadaver. A better result would obviously have been obtained if living full grown larvae had been taken for rearing, or if puparia or adult flies had been collected by policemen in the flat of the woman in question.

The two murdered hitchhiking girls

On 21 August 1971, at 1600, the corpses of two murdered girls who had been hitchhiking were discovered in a sandpit near the town of Hyvinkaa, in southern Finland. The corpses were partially covered by a polyethylene sheet. A cluster of fly eggs was collected from the hair of one of the girls; a fly larvae between 4.5 and 5 mm long was also present in one eye. Four days later, examination of the refrigerated bodies revealed four larvae 5-6 mm long in the eyes of the same girl, and five larvae 2.5-3.5 mm long in the eyes of the other girl.

An attempt was made to rear all the eggs and the larvae to adult flies. The development of the eggs into larvae 4.5-5 mm long (i.e., to the length of the fly larvae observed on 21 August) required one and a half days. Further rearing was only partially successful and a single adult fly of the species *Calliphora vicina* was obtained. Flies of the same species were also obtained from a liver growing-medium placed in a glass container on 28 August 1973 in the place where the dead girls had been found.

Since it had taken one and a half days to obtain a larva 4.5-5 mm long experimentally, it was concluded that the bodies had been in the locality where they were found from 19 August, namely for about two days after the time of death. The suspected murderer, however, had an alibi for 19 August and the following days as well. During the trial the question was raised whether it was possible that the dead girls could have been in the place where they were found on 14 August, as suggested by police investigation. The answer to this question was that, considering the daily temperature from 14 to 19 August (well above 16 degrees Celsius during each day), a large number of big fly larvae should have been found in the corpses. Since this was not the case, one had to draw the conclusion that either the corpses of the girls were not at that place on 14 August or they were completely covered by the polyethylene sheet. From the photograph taken by police immediately after the discovery of the corpses, it seemed possible that the polyethylene sheet at first had covered the girls completely, but later had been partially removed by the wind. The subsequent question was whether fly oviposition could have occurred through possible holes in the polyethylene sheet. The sheet was immediately inspected, but no holes were detected. The results of the police investigation, substantiated by the entomological observations, led to the conviction of the suspected murderer.

Cases from the archives of: Morten Starkeby on Entomology

Greg MacMaster Environmental Forensics

Chapter 17
RESOURCES, WEBSITES, COMPUTER PROGRAMS & SPREADSHEETS

Why re-invent the wheel, right? Since some of us may not have access to data, programs or reference link like others who have been in the business over the years, I decided to make my list and then if you find a site, program or reference that you feel is beneficial to the forensic meteorologist, drop us a line at greg@forensicweatherman.com and well add it to the next revision and credit you for the source!

Resources:

For immediate access to weather logs, try your local National Weather Service Office (NWS) or go online to their website: http://www.crh.noaa.gov/ and find the Climate section. They should have a link to weather over the past year within their CWA.

Daily Weather Maps: Either by subscription or using the online service, it's a way to review past weather in graphic/map format. See http://www.hpc.ncep.noaa.gov/dwm/dwm.shtml for information about this service.

NCDC (National Climactic Data Center): A complete source on all types of weather information, records, graphs, maps, radar, satellite and much more: http://lwf.ncdc.noaa.gov/oa/ncdc.html

NCDC offers free access to a multitude of weather products at this address: http://www4.ncdc.noaa.gov/cgi-win/wwcgi.dll?wwAW~MP~F

If you prefer a more practical approach to data and not have to pay a fee every time, a CD that catalogs megabytes of weather for a single purchase. http://ols.nndc.noaa.gov/plolstore/plsql/olstore.prodlist?category=C&subcatc=01&groupin=CDR

The Climate Visualization system is an interactive graphing tool designed to allow visual browsing of the data available on-line. CLIMVIS simply requires the user to step through the data and graphing feature selection process to visually browse the data. http://www.ncdc.noaa.gov/oa/climate/onlineprod/drought/xmgr.html

Plymouth State Weather Center is site where you can build Contoured 1950-2003 NCEP/NCAR Reanalysis Upper Air Maps. http://vortex.plymouth.edu/reanal-u.html

This product contains historical National Center for Environmental Prediction (NCEP) weather charts that are archived at the National Climatic Data Center. http://ols.ncdc.noaa.gov/cgi-bin/nndc/buyOL-006.cgi?FNC=chart__Ancep_get_chart_htm

For those who wish to utilize those Difax maps of the past, you can go to http://weather.uwyo.edu/difax/index.html and check out the selections available.

You can never have enough resource links: http://www.refdesk.com/weath1.html

You can link directly to an NCDC Weather Observation Station Record for a particular station (or list of stations) using identifiers http://www.ncdc.noaa.gov/oa/climate/linktowcs.html

Yahoo offers an area (Groups) where those with specific skills or knowledge can share it with others. It's a great way to learn more about your job or enhance your skills through others who have previous knowledge of the science. http://groups.yahoo.com/group/forensic_meteorology/

Facebook has many great groups to join and learn: Try 'Environmental Forensics and Its effects on Investigations'

Pave-Cool is a computer program that shows how asphalt cools under changing weather elements. Needless to say, asphalt and the composition can often come into play in legal cases where accumulation of ice and snow is a factor. You can access this program at this addresses; http://www.dot.state.mn.us/app/pavecool/

Bibliography

Benjamin, D. (1993). "Forensic Pharmacology" in R. Saferstein (ed.) Forensic Science Handbook. NJ: Prentice-Hall.

Borg, W.R., and M.D. Gall, Educational Research: An Introduction, New York: Longman Inc., 1983:413–415.

BRE provides research-based consultancy, testing and certification services to customers on buildings, construction, energy, environment, fire and risk world-wide.

Borg and Gall, 413–415.
Clark, S.C., Occupational Research and Assessment, Inc., Big Rapids, Michigan, 1996.

CLP Power Wind/Wave Tunnel Facility : *Central Research Facilities* at the Hong Kong University of Science and Technology.

Combs, D., R.G. Parrish, and R.T. Ing, Death Investigation in the United States and Canada, Atlanta: U.S. Department of Health and Human Services, Public Health Service, Centers for Disease Control and Prevention, 1995.

Compu-Weather - Real-life case examples

MacMaster- How to Plot and Analyze a SKEW T log p Diagram" with Hodograph and rules of thumb cards.

DPLyle,MD 2000–Timely Death

Environmental Adaptation Research Group, Atmospheric Environment
Services Downsview, Ontario, Canada

Fernando Caracena NOAA, Forecast Systems Laboratory, Boulder, Colorado . Microburst study with Theodore Fujita

Feinman, Law 101: Everything You Need to Know About the American Legal System, New York: Oxford University Press,

2000. ISBN: 019 513 265 3

Haggard–2003–Definition of a Forensic Meteorologist

Hanzlick, R., "Coroner Training Needs: A Numeric and Geographic Analysis," Journal of the American Medical Association 276 (1996):1775–1778.

Jay Dix and Michael Graham - (Photo—Time of Death, Decomposition and Identification, an Atlas)

Jentzen, J.M., S.C. Clark, and M.F. Ernst, "Medicolegal Death Investigator Pre-Employment Test Development," American

Journal of Forensic Medicine and Pathology 17 (1996):112-16.
J.S. Clauzel Flight Safety-DACO-1981

Karen Holmes Balestreri, Pendleton & Potocki as presented at the AMS Conference of Forensic Meteorologist. Jan/2005

Klaasen, C. (1996). "Principles of Toxicology and Treatment of Poisoning" in J. Hardman et al., Goodman and Gilman's The Pharmacological Basis of Therapeutics. NY: McGraw-Hill.

Levine, A. (1993). "Forensic Toxicology" Journal of Analytical Chemistry 65: 272–76.

Lowry, W. & J. Garriott. (1979). Forensic Toxicology: Controlled Substances and Dangerous Drugs. NY: Plenum.

MacMaster, Greg: Real-life cases examples

Moenssens, A.A.; Inbau, F.E.; Starrs, J.E. (1986). Scientific Evidence in Civil and Criminal Cases. NY: The Foundation Press.

Morten Starkeby - Cases from the archives

NTSB—Aviation Accident Investigation information referenced from: *http://www.nws.noaa.gov/om/forensic.shtml*

Paul Mason Consulting. RFID - Retail Sales

Randall B. Christison Attorney at Law as presented at the American Meteorological Society Conference of Forensic Meteorology

Saferstein, R. (1998). Criminalistics: An Introduction to Forensic

Science. NJ: Prentice-Hall.

Schubert, Grillot's Introduction to Law and the Legal System, 6th Ed., Boston: Houghton Mifflin Co., 1996. ASIN: 039 574 666 3

The Ohio State University, Center on Education and Training for Employment, DACUM, 1996.

Thorntons, May 2003 -Trading Statement by
Benjamin, D. (1993). "Forensic Pharmacology" in R. Saferstein (ed.) *Forensic Science Handbook*. NJ: Prentice-Hall.

Klaasen, C. (1996). "Principles of Toxicology and Treatment of Poisoning" in J. Hardman et al., *Goodman and Gilman's The Pharmacological Basis of Therapeutics*. NY: McGraw-Hill.

Levine, A. (1993). "Forensic Toxicology" *Journal of Analytical Chemistry* 65: 272–76.

Lowry, W. & J. Garriott. (1979). *Forensic Toxicology: Controlled Substances and Dangerous Drugs*. NY: Plenum.

Moenssens, A.A.; Inbau, F.E.; Starrs, J.E. (1986). Scientific Evidence in Civil and Criminal Cases. NY: The Foundation Press.
Saferstein, R. (1998). *Criminalistics: An Introduction to Forensic*

Science. NJ: Prentice-Hall.

Any other references left out were purely unintentional.